LOW-VOLTAGE LOW-POWER CMOS CURRENT CONVEYORS

Low-Voltage Low-Power CMOS Current Conveyors

by

Giuseppe Ferri
University of L'Aquila, Italy

and

Nicola C. Guerrini
University of L'Aquila, Italy

KLUWER ACADEMIC PUBLISHERS
BOSTON / DORDRECHT / LONDON

A C.I.P. Catalogue record for this book is available from the Library of Congress.

ISBN 1-4020-7486-7

Published by Kluwer Academic Publishers,
P.O. Box 17, 3300 AA Dordrecht, The Netherlands.

Sold and distributed in North, Central and South America
by Kluwer Academic Publishers,
101 Philip Drive, Norwell, MA 02061, U.S.A.

In all other countries, sold and distributed
by Kluwer Academic Publishers,
P.O. Box 322, 3300 AH Dordrecht, The Netherlands.

Printed on acid-free paper

Printed in the Netherlands.

Abstract

Research in analog integrated circuits has recently gone in the direction of low-voltage (LV), low-power (LP) design, especially in the environment of portable systems where a low supply voltage, given by a single-cell battery, is used. These LV circuits have to show also a reduced power consumption to maintain a longer battery lifetime. In this area, traditional voltage-mode techniques are going to be substituted by the current-mode approach, which has the recognized advantage to overcome the gain-bandwidth product limitation, typical of operational amplifiers. Then, they do not require high voltage gains and have good performance in terms of speed, bandwidth and accuracy. Inside the current-mode architectures, the current-conveyor (CCII) can be considered the basic circuit block because all the active devices can be made of a suitable connection of one or two CCIIs. CCII is particularly attractive in portable systems, where LV LP constraints have to be taken into account. In fact, it suffers less from the limitation of low current utilisation, while showing full dynamic characteristics at reduced supplies (especially CMOS version) and good high frequency performance. Recent advances in integrated circuit technology have also highlighted the usefulness of CCII solutions in a large number of signal processing applications.

The outline of *Low voltage low power CMOS current conveyors* is the following. In the first chapter, the authors talk about the current-mode approach and a brief history of the first and second generation CC. Then, the second generation current-conveyor (CCII) will be considered as a building block in the main active feedback devices and in the implementation of simple analog functions, as an alternative to OA. In the second chapter, the design and characteristics of CCII topologies are described, together with a further look into CCII modern solutions and future trends. Chapter 3 deals with low voltage low power LV LP CCII implementations and new considerations about CCII noise and offset. In Chapter 4 the CCII evolution towards differential and generalized topologies will be considered. Chapter 5 deals about old and new CCII applications in some basic analog functions such as filters, impedance simulators and converters, oscillators, etc.. In the Appendix, there is also an experimental session, where on-chip measurements can be compared with theory and simulations.

Low voltage low power CMOS current conveyors is a valuable reference source for current-mode and CCII analog integrated circuit designers and can be considered also a suitable text for advance courses on microelectronics.

Table of Contents

CHAPTER I

INTRODUCTION

The development of VLSI technology, together with the request of a larger number of elements on a single chip, has led to an improved interest in analog circuit design, especially for what concerns integrated circuits. The main aim of analog integrated circuits (ICs) is to satisfy circuit specifications through circuit architectures with the required performance. They can be used either as "stand-alone" topologies or connected to the digital part to implement mixed analog-digital functions, utilised in a wide field of applications. Even if numerous researchers have predicted a reduced utilisation of analog architectures and an increased development of the digital counterpart, analog circuitry continues to be necessary. In fact, analog circuits are needed in many VLSI systems such as filters, D/A and A/D converters, voltage comparators, current and voltage amplifiers, etc.. Moreover, new applications continue to appear where new analog topologies have to be designed to ensure the trade-off between speed and power requirements. Finally, the recent trend towards miniaturized circuits has given a strong and decisive boost towards the design of low-voltage low-power (LV LP) analog integrated circuits, which are widely utilized in portable-system applications [1,2,3]. This has led to implement new design circuit strategies in low-cost CMOS technology.

Since the beginning of electronics, the need of new active devices has always been very important. It has driven to the birth of transistors which have been used, then, in amplifiers, impedance converters, filters, etc.. In particular, the voltage operational amplifier (OA) has rapidly become the main analog block and has dominated the market since the advent of the first analog integrated circuits. Nowadays, the situation is changing because there is a new impulse towards the so called *current-mode* circuits [4,5,6,7], which are able to overcome the limitation of a constant gain-bandwidth product [4,8,9] and the trade-off between speed and bandwidth, so that performance is improved in terms of low-voltage characteristics and of slew-rate and bandwidth.

The outline of the book is the following: firstly, the authors will talk about a brief history of the first and second generation CC. Then, the second generation current-conveyor (CCII) will be considered as a building block in the main active feedback devices and in the implementation of simple analog functions, as an alternative to OA. In the next chapters, the design of CCII topologies will be considered together with a further look into CCII modern solutions and future trends, in particular low offset and low noise topologies. That is why, the authors will then describe LV LP CCII implementations, their evolution towards differential and generalized topologies, and new possible CCII applications in some basic analog functions such as filters, impedance simulators and converters, oscillators, etc.. Some of these implementations have been fabricated in a standard CMOS technology. On chip measurements will be finally reported.

1.1 THE CURRENT-MODE APPROACH: BRIEF HISTORY OF CURRENT CONVEYORS.

1.1.1 The current-mode approach.

In analog circuit design, there is often a large request for amplifiers with specific current performance for signal processing.
The *current-mode* approach [4,6], which considers the information flowing on time-varying currents, proposes a new way to "see" integrated circuits even if sometimes there is nothing new, except that we are revisiting old circuits towards different and more elegant solutions for many circuit problems.
Current-mode techniques are characterised by signals as typically processed in the current domain. Current-mode circuits have some recognised advantages: firstly, they do not require a high voltage gain, so high performance amplifiers are not needed. Then, they do not need high precision passive components, so they can be designed almost entirely with transistors. This makes the current-mode circuits compatible with typical digital processes. Finally, they show high performance in terms of speed, bandwidth and accuracy.
The current-mode approach is also powerful if we consider that all the analog IC functions, which traditionally were been designed in the *voltage-mode*, can be also implemented in current-mode.

A well-known current-mode circuit is the Current-Feedback Operational Amplifier (CFOA) [10,11,12,13,14]. This circuit, if compared to the traditional voltage OA, shows a constant bandwidth with respect to the closed-loop gain and a very high slew-rate. This makes this circuit of primary importance in the design of modern LV LP ICs.

The first stage of CFOA is the current-conveyor (CC), which is the main subject of this book. As a matter of facts, CC can be considered the basic current-mode building block because all the active devices can be made of a suitable connection of one or two CCIIs. It will be particularly attractive in the environment of portable systems where a low supply voltage, given by a single-cell battery, is used. These LV circuits have to show also a reduced power consumption to maintain a longer battery lifetime. This implies a reduction of the biasing currents in the amplifier stages, with consequent reduction in some amplifier performance. The current-mode approach suffers less from this limitation, while showing full dynamic characteristics also at reduced supply levels and good high-frequency performance.

Moreover, recent advances in the technologies of ICs have highlighted the usefulness of the attractive and elegant current-mode circuit solutions in a large number of signal processing applications.

1.1.2 Brief history of first and second generation current conveyors

The current conveyor (CC) is a basic block that can be implemented in analog circuit design using a like-OA approach; it also represents an effective alternative to the same OA for designers. This is mainly due to the fact that both practical current conveyors and OAs are marked by characteristics that are very close to the ideal ones.

Sedra and Smith introduced the current conveyors in 1968 [15,16] but their real advantages and innovative impact were not immediately clear at that time. In fact, at the same time, electronic companies started to put their main efforts in the fabrication of monolithic OAs; as a consequence, the relevant value of the new invention was partially overshadowed.

Only in recent years, with the growing diffusion of the current-mode approach as a way to design LV LP circuits, current conveyors have gained an increased popularity.

The original example presented by Sedra and Smith in 1968 was generically named by the authors "current conveyor". The first block was identified as "first generation current conveyor", or CCI, only when its evolved topology was called "second generation current conveyor", or CCII, in 1970 [16,17].
CCI is a three-terminal device, schematically represented in figure 1.1.

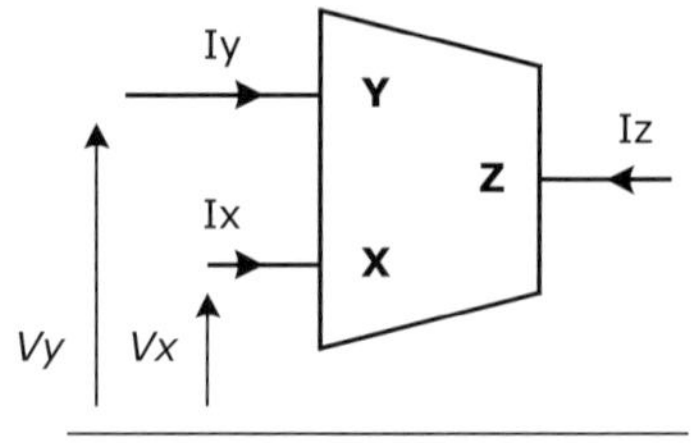

Figure 1.1 – CCI block representation

It operates as follows: if a voltage is applied to Y node, the same voltage will appear at X node, while the opposite happens to currents. In fact, the current flowing at Y node is equal to the one flowing at X node; this current is "CONVEYED" to the output Z node, too.

$$\begin{bmatrix} Iy \\ Vx \\ Iz \end{bmatrix} = \begin{bmatrix} 0 & 1 & 0 \\ 1 & 0 & 0 \\ 0 & \pm 1 & 0 \end{bmatrix} \begin{bmatrix} Vy \\ Ix \\ Vz \end{bmatrix}$$

CCI Node	Impedance level
X	Low (ideally 0)
Y	Low (ideally 0)
Z	High (ideally ∞)

Figure 1.2 – CCI main characteristics

Current at Z node can flow in the direction of Ix or in the opposite one. In the matrix description reported in figure 1.2, we assume that sign + stays for currents flowing in the same direction, while sign – stays for the opposite situation, considering CCI as reference. In the first case we have a "positive CCI" (named CCI+), in the second case a "negative CCI" (or CCI-). X and Y nodes have a low impedance level, ideally zero, whereas Z node shows a high impedance level, ideally infinite (figure 1.2).
The previous description clearly points out how the voltage at X node is independent of the current flowing at the same node. Similarly, the current flowing at Y node is not related to the potential applied in any way.

It is possible to represent CCI using the nullator-norator (nullor) formalism (see Appendix A), as in figure 1.3 [16],

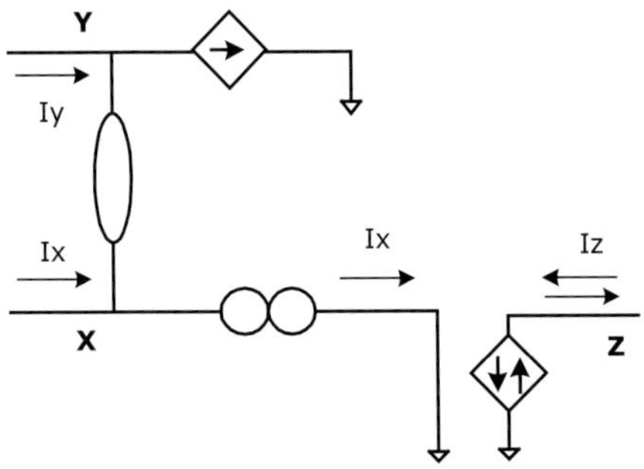

Figure 1.3 – Nullator-norator CCI model

while figure 1.4 reproduces a possible practical realisation for a class A CCI at transistor level [16].

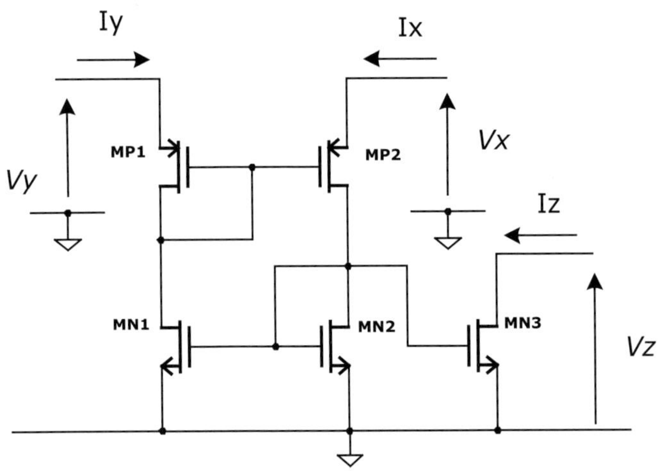

Figure 1.4 – Class A CCI

The circuit topology can be implemented either in bipolar or CMOS technology, but, nowadays, the latter is much more used than the former. Its main drawback is represented by the fact that it operates in class A. To improve the circuit clearness, the biasing circuits have not been considered.

The operating mode of the circuit shown in figure 1.4 is easily derived from the topology shown. MP1 and MP2 perform the voltage following action between Y and X nodes, while the current mirror, formed by MN1 and MN2, provides a current Iy equal to that flowing from X node. Through the use of MN3, the same current is "conveyed" to the high-impedance current output node Z.
Starting from this topology, a simple class AB CCI can be designed, as shown in figure 1.5 [16,18].

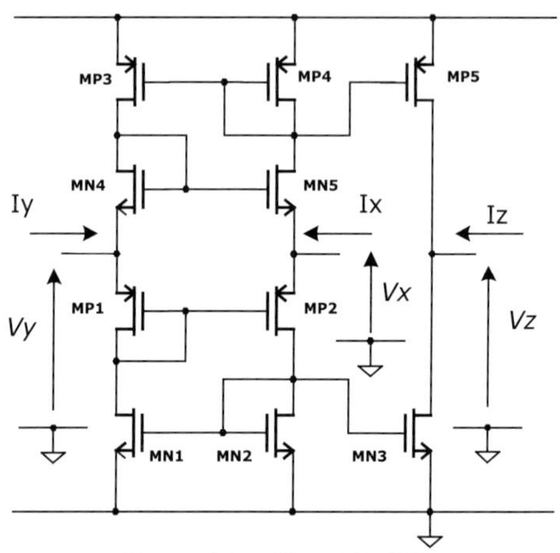

Figure 1.5 – Class AB CCI

Another possible topology for the first generation current conveyor, which includes bias stabilisation, is shown in figure 1.6 [19]. In fact, the quiescent current flowing in the X-node branch is very sensitive to supply voltage and transistors mismatch. Also, current mirrors MN1-MN2 and MP3-MP4 may present a gain lower than unity. This leads to a bad control of the biasing current with consequent bad circuit performance. A simple but effective way to reduce this problem consists in feeding back, to the Y node branch, only a fixed fraction of X node current. The remaining part will be provided by current generators named Iss.

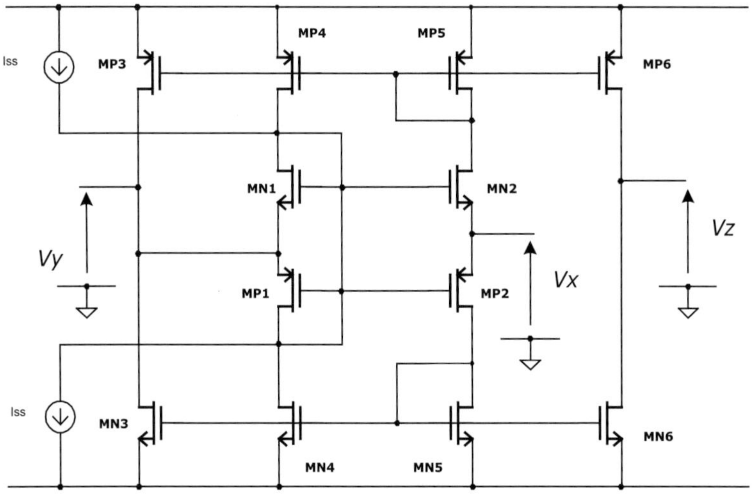

Figure 1.6 – Current conveyor topology with bias stabilisation

When CCI was firstly introduced, it was employed as a new building block in the design of simple analog signal processing circuits. For example, we have implemented topologies as: V-I and I-V converters, negative impedance converters. They are shown in figures 1.7,1.8,1.9.

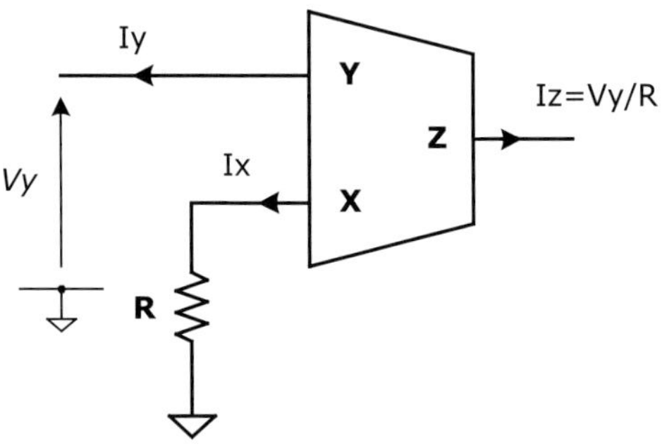

Figure 1.7 – CCI-based V to I converter

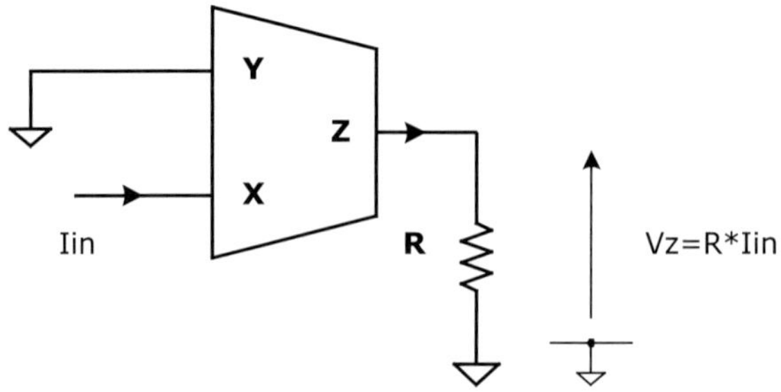

Figure 1.8 – CCI-based I to V converter

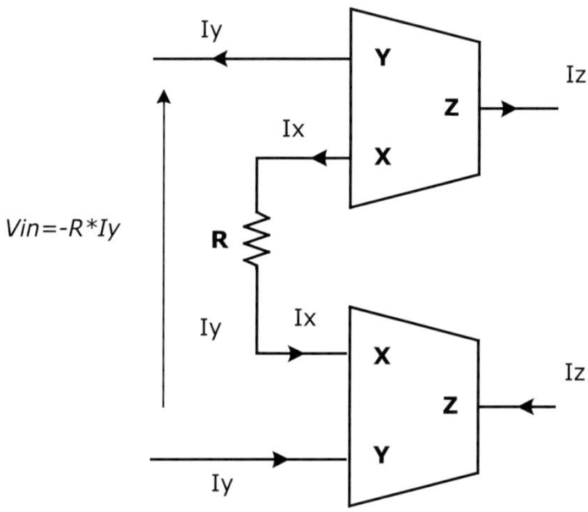

Figure 1.9 – CCI-based differential negative impedance converter

The design of circuits based on CCI can, in some cases, turn out to be quite problematic, since the current flows in all the block terminals. This is perhaps the greater limit of the CCI device, which reduces its flexibility and versatility.

As stated before, the current conveyor success came only when CCII was introduced, two years later than CCI. Basically, there is only a little difference between the two blocks, but in the practical applications CCII has shown to be much more versatile and helpful than CCI.

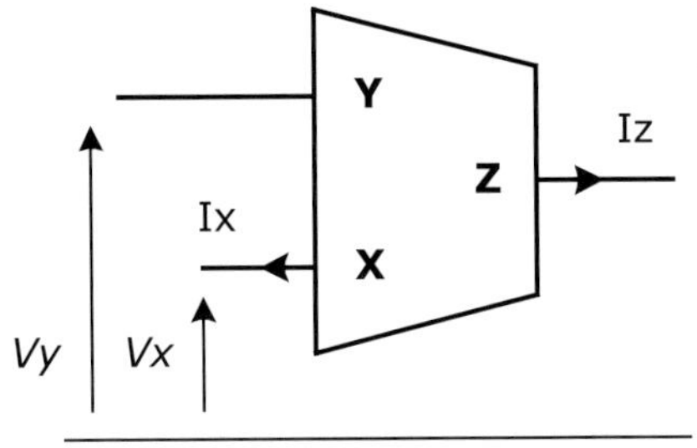

Figure 1.10 - CCII block representation

CCII is topologically very similar to its predecessor (see figure 1.10) [16,17]. The electrical characteristics of the new block are reported in figure 1.11.

$$\begin{bmatrix} Iy \\ Vx \\ Iz \end{bmatrix} = \begin{bmatrix} 0 & 0 & 0 \\ 1 & 0 & 0 \\ 0 & 1 & 0 \end{bmatrix} \begin{bmatrix} Vy \\ Ix \\ Vz \end{bmatrix}$$

CCII Node	Impedance level
X	Low (ideally 0)
Y	High (ideally ∞)
Z	High (ideally ∞)

Figure 1.11 – CCII main characteristics

Compared to the previous version, the innovation of CCII is represented by the absence of current in the Y node, owing to its high impedance (ideally infinite). From the nullor point of view, CCII can be easily represented, as shown in figure 1.12, where only a simplified scheme is depicted [16,20].

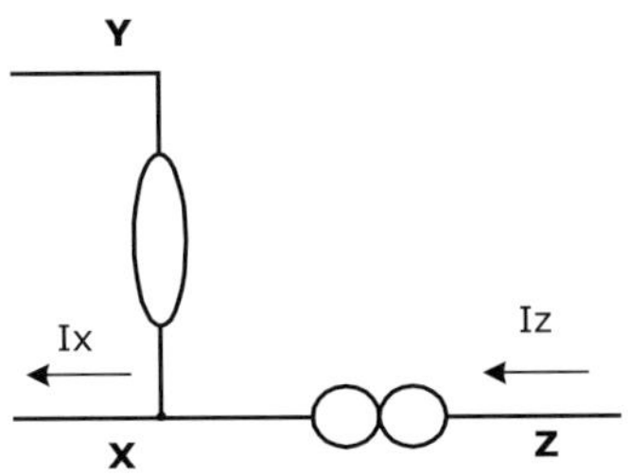

Figure 1.12 – Nullator-norator CCII simplified model

Using this model, currents flowing at X and Z nodes are equal in magnitude but opposite in direction, assuming CCII as their reference. This means that if Ix flows out from X node, Iz flows into Z node and vice versa.
On the other hand, real CCII implementations can lead to two different situations, as for CCI. In figure 1.12 (currents flowing in opposite directions from the CCII point of view), the block is called negative CCII (CCII-) while, when Ix and Iz flow in the same direction, we have a positive CCII (CCII+). After this statement, it is clear that figure 1.10 represents a CCII+.
A more complete nullator-norator model for CCII, which takes into account the two possibilities described above, is represented in figure1.13 [16].

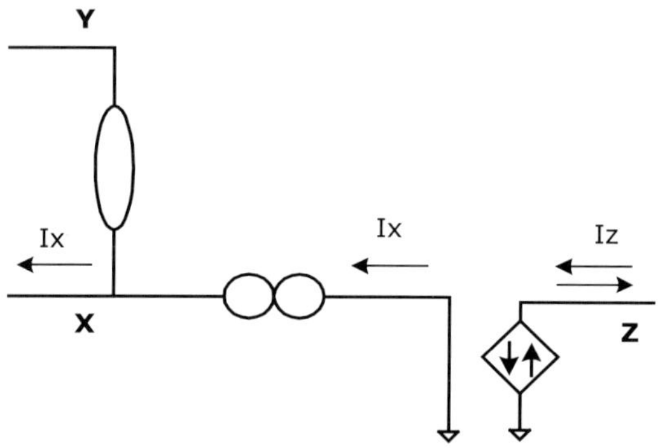

Figure 1.13 – Nullator-norator CCII complete scheme

The two cases of negative and positive CCII are also included in the more complete matrix description of figure 1.14.

$$\begin{bmatrix} Iy \\ Vx \\ Iz \end{bmatrix} = \begin{bmatrix} 0 & 0 & 0 \\ 1 & 0 & 0 \\ 0 & \pm 1 & 0 \end{bmatrix} \begin{bmatrix} Vy \\ Ix \\ Vz \end{bmatrix}$$

Figure 1.14 – CCII complete matrix description

Many different solutions, some of which will be illustrated and explained in detail in chapters 2 and 3, have been studied and proposed in the literature to implement CCII.

Nevertheless, CCII success does not rely on a particular circuit solution but on its general aptitude of being so easily used in analog processing circuits, like – and sometimes with better results – the well-known OA. In the figures 1.15-1.21, simple applications with CCIIs, typically implemented with OAs (VCVS, VCCS, CCCS, CCVS, current amplifier, current differentiator and integrator), are shown [16].

In the following applications, negative and positive CCIIs are employed. The different cases can be distinguished by looking at the arrows showing the current flow.

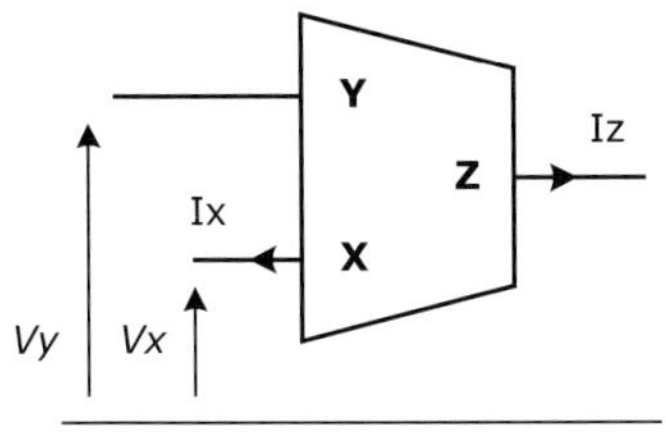

Figure 1.15 – CCII-based voltage controlled voltage source (VCVS)

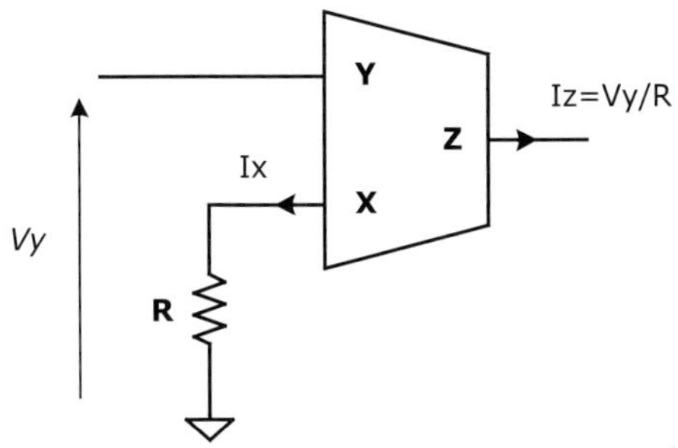

Figure 1.16 – CCII-based voltage controlled current source (VCCS)

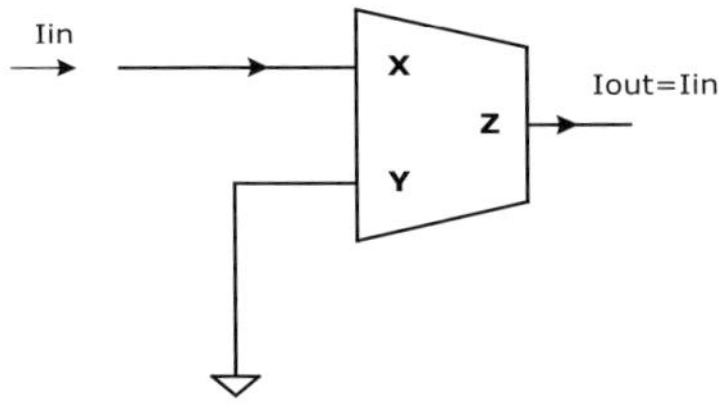

Figure 1.17 – CCII-based current controlled current source (CCCS)

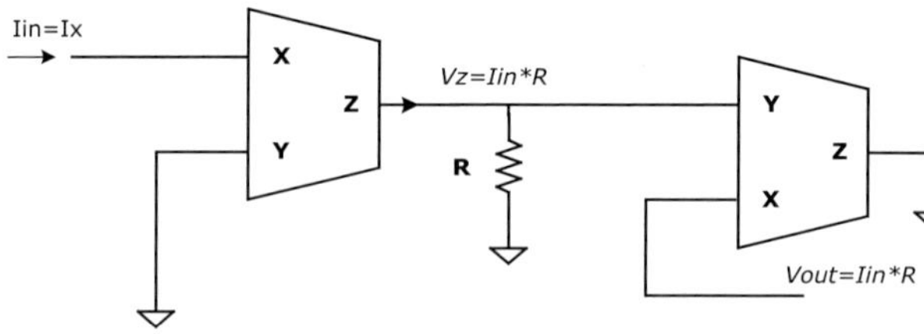

Figure 1.18 – CCII-based current controlled voltage source (CCVS)

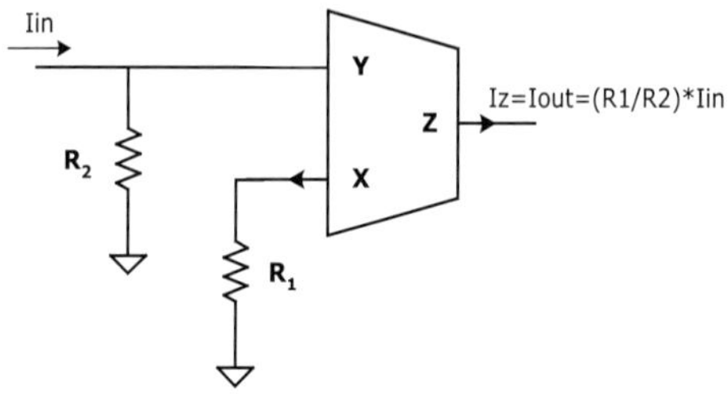

Figure 1.19 –CCII-based current amplifier

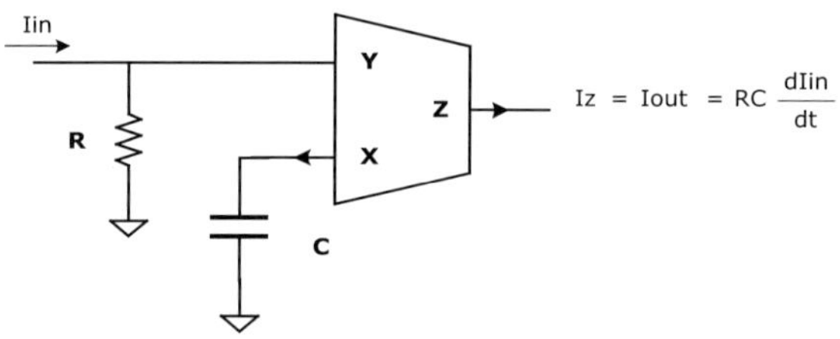

Figure 1.20 – CCII-based current differentiator

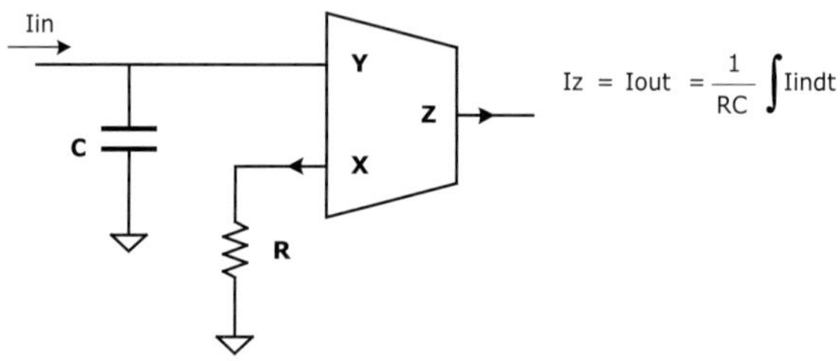

Figure 1.21 - CCII-based current integrator

1.2 THE SECOND GENERATION CURRENT CONVEYOR (CCII) AS BUILDING BLOCK.

1.2.1 The nullor approach.

The concept of an ideal active device has been introduced in electronic theory world in 1954 by Tellegen [21]. Ten years later, this ideal amplifier was represented by the nullator-norator (see Appendix A) formalism by Carlin, who introduced the nullor concept [22], reported in figure 1.22.

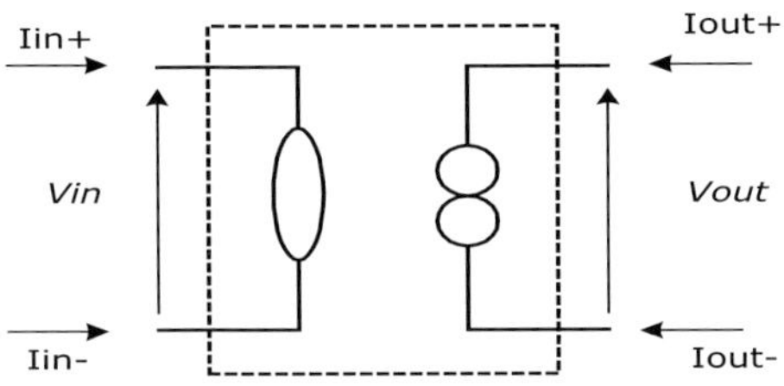

Figure 1.22 – Nullor symbol

The nullor can be considered as the pre-eminently ideal amplifier and it is used as starting point in almost every theoretical approach to the amplifiers. Active elements, such as OAs or CCIIs, can be regarded as partial realisation of nullors. In fact, ideal OAs and CCIIs are simply obtained grounding one of the norator and one of the nullator terminals, respectively, as shown in detail in Appendix A [20]. (A different nullator-based model for the CCII, maybe more intuitive, will be shown in figure 1.24).

Due to their direct derivation from nullor, OA and CCII- propose themselves, with the same importance, as fundamental active devices.

Nullor has undefined impedance levels, so its transfer properties are well defined only if an external network performs a feedback between output and input ports. Basically, in analog design, the following four fundamental active blocks are considered.

- Voltage Controlled Voltage Source (VCVS) = voltage amplifier
- Voltage Controlled Current Source (VCCS) = transconductance amplifier
- Current Controlled Voltage Source (CCVS) = transresistance amplifier
- Current Controlled Current Source (CCCS) = current amplifier

For all these gain elements, it is possible to implement a nullor model [20] as reported in figure 1.23.

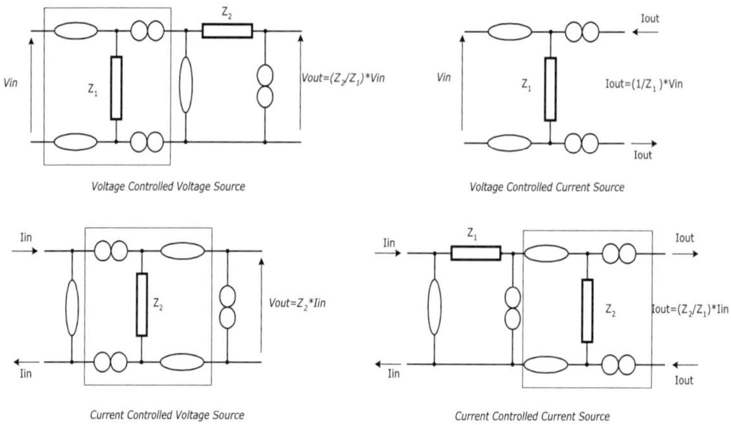

Figure 1.23 – VCVS, VCCS, CCVS and CCCS nullator-norator models

The dotted line squares highlight the common part of the four basic gain blocks. This common part is basically a VCCS and can be viewed in a different way simply removing Z impedance, as in figure 1.24, which shows a sort of differential nullator-norator model for CCII. In fact, the differential voltage applied between Y and Y' nodes is equal to the one between X and X' nodes.

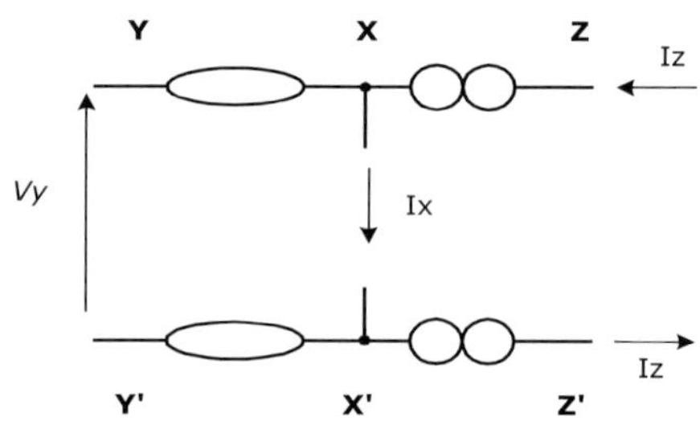

Figure 1.24 – CCII nullator-norator model

Thanks to the presence of the two norators, the current flowing between Z and Z' nodes is opposite to that flowing between X and X' nodes, which is a typical characteristic of CCII-. Consequently, in figure 1.23 we have represented a set of four differential nullator-norator models where the CCII- is a fundamental internal element. This gives a confirmation to the fact that CCII is a key element in the design of active devices.

1.2.2 The feedback amplifier approach.

The nullor, introduced in the previous paragraph, is only an ideal element used in the amplifier theory, but it has no meaning if used without an external (feedback) network, as reported before and in appendix A, owing it to the undefined impedance levels and transfer characteristics [13,20,23].
Moreover, practical circuit implementations have to deal with non ideal features of the basic gain elements, also in terms of input and output impedance levels. It is well known how the main characteristics of a given amplifier can be usually improved by means of feedback networks. Four different feedback configurations are possible, as depicted in figure 1.25 as block schemes. In figure 1.26, possible implementations of the feedback configurations, at circuit level and considering a generic gain element, are shown [8,23,24,25,26].

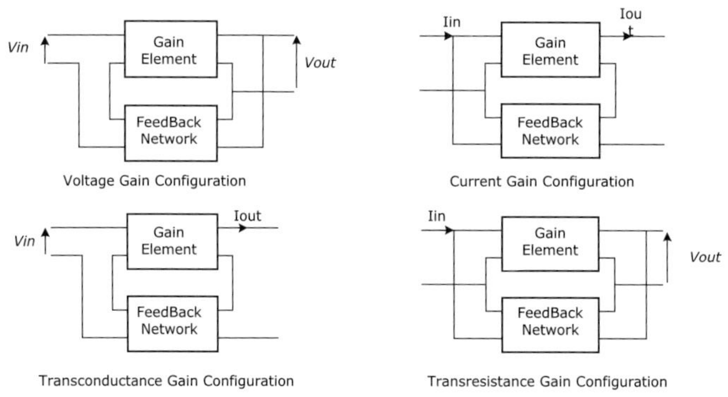

Figure 1.25 – Feedback configuration block schemes

Considering the four kinds of gain elements (VCVS, VCCS, CCVS and CCCS) in each of the four different feedback networks, it follows that sixteen combinations are possible. In table 1.1, the main characteristics of the basic gain elements and feedback network are reported, where A is the open loop gain and β the feedback gain [8].

Voltage Gain Configuration					
Gain Element	**Input Impedance**	**Output Impedance**	**BandWidth**	**Input Impedance with Feedback**	**Output Impedance with Feedback**
Voltage Amplifier	High	Low	GBW Constant	Enhanced by (1+Aβ)	Lowered by (1+Aβ)
Current Amplifier	Low	High	Constant	Enhanced by (1+Aβ)	Lowered by (1+Aβ)
Transresistance Amplifier	Low	Low	Potentially Constant	Enhanced by (1+Aβ)	Lowered by (1+Aβ)
Transconductance Amplifier	High	High	Potentially Constant	Enhanced by (1+Aβ)	Lowered by (1+Aβ)
Current Gain Configuration					
Gain Element	**Input Impedance**	**Output Impedance**	**BandWidth**	**Input Impedance with Feedback**	**Output Impedance with Feedback**
Voltage Amplifier	High	Low	Constant	Lowered by (1+Aβ)	Enhanced by (1+Aβ)
Current Amplifier	Low	High	GBW Constant	Lowered by (1+Aβ)	Enhanced by (1+Aβ)
Transresistance Amplifier	Low	Low	Potentially Constant	Lowered by (1+Aβ)	Enhanced by (1+Aβ)
Transconductance Amplifier	High	High	Potentially Constant	Lowered by (1+Aβ)	Enhanced by (1+Aβ)
Transresistance Gain Configuration					
Gain Element	**Input Impedance**	**Output Impedance**	**BandWidth**	**Input Impedance with Feedback**	**Output Impedance with Feedback**
Voltage Amplifier	High	Low	Constant	Lowered by (1+Aβ)	Lowered by (1+Aβ)
Current Amplifier	Low	High	Constant	Lowered by (1+Aβ)	Lowered by (1+Aβ)
Transresistance Amplifier	Low	Low	GBW Constant	Lowered by (1+Aβ)	Lowered by (1+Aβ)
Transconductance Amplifier	High	High	Potentially Constant	Lowered by (1+Aβ)	Lowered by (1+Aβ)
Transconductance Gain Configuration					
Gain Element	**Input Impedance**	**Output Impedance**	**BandWidth**	**Input Impedance with Feedback**	**Output Impedance with Feedback**
Voltage Amplifier	High	Low	Constant	Enhanced by (1+Aβ)	Enhanced by (1+Aβ)
Current Amplifier	Low	High	Constant	Enhanced by (1+Aβ)	Enhanced by (1+Aβ)
Transresistance Amplifier	Low	Low	Potentially Constant	Enhanced by (1+Aβ)	Enhanced by (1+Aβ)
Transconductance Amplifier	High	High	GBW Constant	Enhanced by (1+Aβ)	Enhanced by (1+Aβ)

Table 1.1 – Amplifiers and feedback characteristics

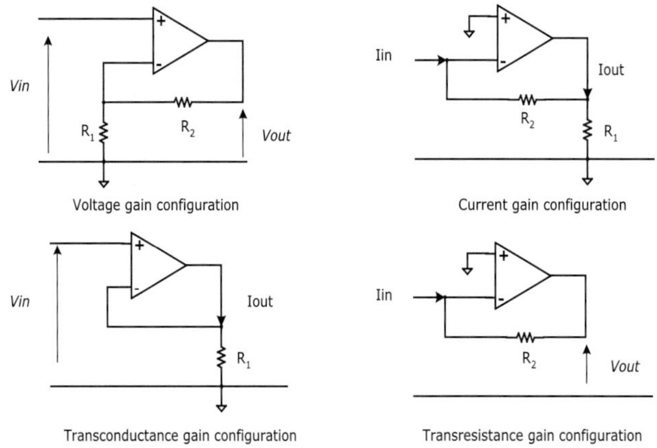

Figure 1.26 – Possible feedback configurations

A strongly desired feature for a feedback network is its capability to give a network having, as a result, independent characteristics of the employed amplifier. In other words, a good feedback topology has to be able to give the best performance independently of the gain-element characteristics. For example, we can consider a current amplifier with the current gain configuration applied on.

Voltage Gain Configuration					
Gain Element	**Input Impedance**	**Output Impedance**	**BandWidth**	**Input Impedance with Feedback**	**Output Impedance with Feedback**
Voltage Amplifier	High	Low	GBW Constant	Enhanced by (1+Aβ)	Lowered by (1+Aβ)
Current Gain Configuration					
Gain Element	**Input Impedance**	**Output Impedance**	**BandWidth**	**Input Impedance with Feedback**	**Output Impedance with Feedback**
Current Amplifier	Low	High	GBW Constant	Lowered by (1+Aβ)	Enhanced by (1+Aβ)
Transresistance Gain Configuration					
Gain Element	**Input Impedance**	**Output Impedance**	**BandWidth**	**Input Impedance with Feedback**	**Output Impedance with Feedback**
Transresistance Amplifier	Low	Low	GBW Constant	Lowered by (1+Aβ)	Lowered by (1+Aβ)
Transconductance Gain Configuration					
Gain Element	**Input Impedance**	**Output Impedance**	**BandWidth**	**Input Impedance with Feedback**	**Output Impedance with Feedback**
Transconductance Amplifier	High	High	GBW Constant	Enhanced by (1+Aβ)	Enhanced by (1+Aβ)

Table 1.2 – Enhanced feedback combinations

The external network added to the gain element improves some of the current amplifier characteristics, performing lower input and higher output impedances, as from table 1.1 [8]. The concept of "enhanced" feedback comes directly from these considerations. Among the sixteen feedback combinations, four of them can be identified as "preferred" ones [25]. The complete set of "enhanced" combinations, taken from table 1.1, is shown in table 1.2.
They have been chosen from those presented in table 1.1 considering the feedback network that "improves" the impedance levels of the overall circuit.
Even if these combinations are named also "preferred", they have a drawback: their gain-bandwidth product (GBW) is constant. This is the case, for example, of voltage amplifiers compensated with a single dominant pole which have a very low cut-off frequency. A higher bandwidth can be obtained either reducing the voltage gain or, alternatively, *implementing different feedback solutions.* In fact, looking at table 1.1, it is possible to see that it is not unusual to obtain a feedback amplifier with constant bandwidth.

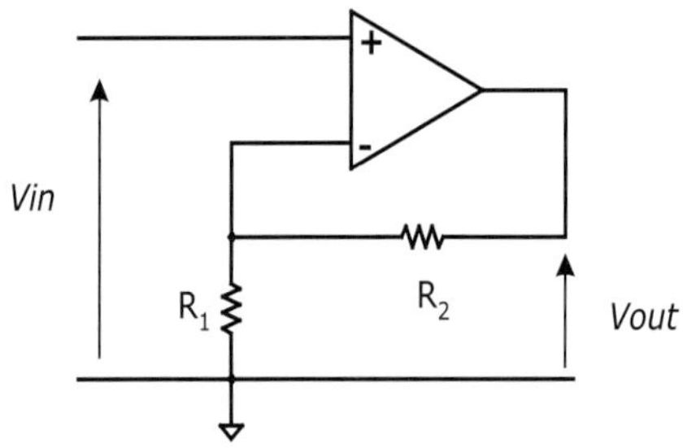

Figure 1.27 – Voltage gain device in a voltage gain feedback configuration

In fact, if we consider a voltage gain element in a voltage amplifier feedback network, as in figure 1.27, we can write:

$$\frac{V_{OUT}}{V_{IN}} = \frac{A}{1 + A\dfrac{R_1}{R_1 + R_2}} = \frac{A}{1 + A\beta}, \quad \text{being} \quad \beta = \frac{R_1}{R_1 + R_2}; \tag{1.1}$$

and, considering a single pole amplifier, that is:

$$A(s) = \frac{A_0}{1 + \dfrac{s}{\omega_P}}, \quad \text{we have:} \quad \frac{V_{OUT}}{V_{IN}} = \frac{\dfrac{A_0}{1 + A_0\beta}}{1 + \dfrac{s}{\omega_P(1 + A_0\beta)}} \tag{1.2}$$

The pole of the frequency response is increased by the same factor to which the gain is reduced. From the previous formulas it is clear that gain-bandwidth product, for such a combination of gain element and feedback network, is constant.
Let us consider now a different case, a voltage gain device in a current gain configuration, shown in figure 1.28.

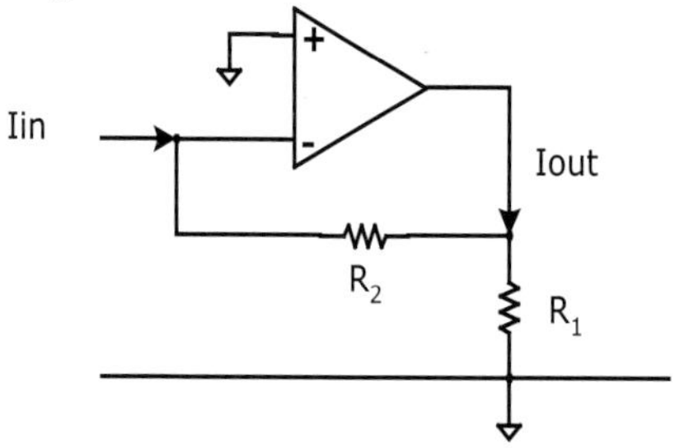

Figure 1.28 – Voltage gain device in a current gain feedback configuration

This is a different situation, for which we can determine the current gain K (in feedback). It can be written that:

$$I_{OUT} = KI_{IN} = K\frac{V_{IN} - V_{OUT}}{R_2} \tag{1.3}$$

since $I_{OUT} + I_{IN} = \frac{V_{OUT}}{R_1}$ and $V_{OUT} = -A\,V_{IN}$, we have: $K = -(1 + \frac{R_2}{R_1}\frac{A}{A+1}) \approx -(1 + \frac{R_2}{R_1})$ (1.4)

If A is a single pole amplifier, we can write:

$$A(s) = \frac{A_0}{1 + \frac{s}{\omega_P}}, \text{ then: } K = -(1 + \frac{R_2}{R_1}\frac{\frac{A_0}{1+\frac{s}{\omega_P}}}{\frac{A_0}{1+\frac{s}{\omega_P}} + 1}) = -(1 + \frac{R_2}{R_1}\frac{\frac{A_0}{A_0+1}}{1 + \frac{s}{\omega_P(A_0+1)}}) \approx -(1 + \frac{\frac{R_2}{R_1}}{1 + \frac{s}{\omega_P(A_0+1)}}) \tag{1.5}$$

From eq.s (1.2) and (1.5), we can say that the current gain value can be obtained choosing R_1 and R_2, and that the feedback has increased the bandwidth decreasing the gain in the voltage configuration (see figure 1.29), while it has increased the bandwidth without affecting the gain in the current amplifier configuration (see figure 1.30).

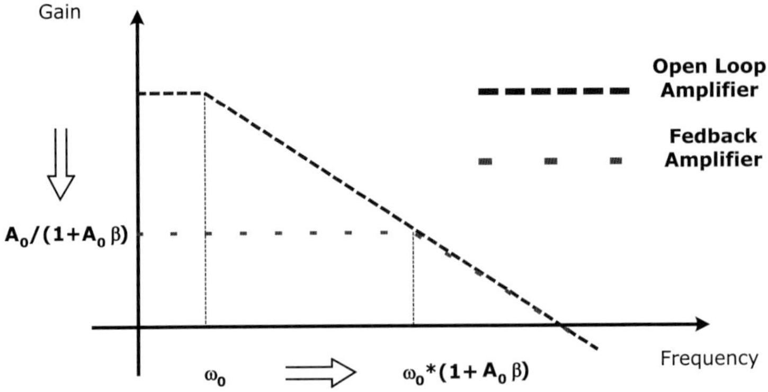

Figure 1.29 – Feedback effects on voltage amplifier frequency response

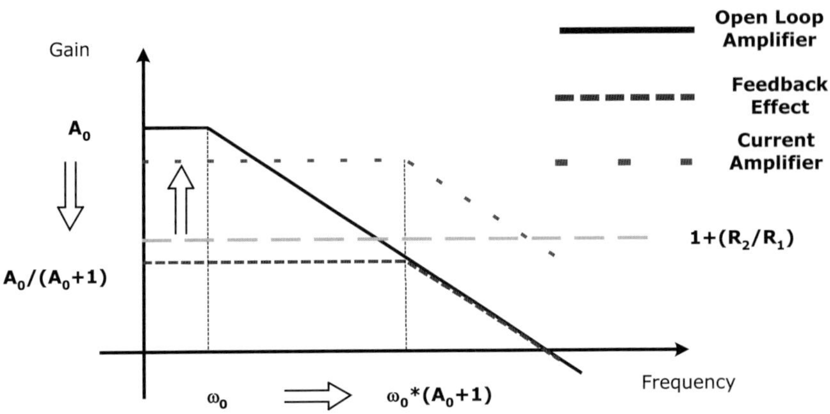

Figure 1.30 – Feedback effects on current amplifier frequency response

In the search for better active elements, the LV LP philosophy [1,2] can play an important role. In fact, due to the restricted voltage swing allowed, it is not always possible to connect two port networks in series, so both current sensing and voltage comparing may not be the best choice for feedback. Then, for LV LP applications, only the transresistance feedback configuration, which considers voltage sensing and current comparing, seems to be utilizable or, at least, the more appropriate [27].

From the above considerations, it comes that a transconductance amplifier (figure 1.31) in a transresistance feedback configuration (figure 1.32) [13,28] can be the best feedback topology, because in this configuration this kind of amplifier does not have the limitation of constant GBW, as it results in table 1.1. The other two possible gain elements (voltage and current amplifiers), which do not limit the GBW product, are not taken into consideration because they are less suitable for the implementation of a general purpose block, as proved later on.
In fact, since the transresistance feedback has the effect to decrease the input and the output impedance levels (at terminals 2 and 4 in the circuit of figure 1.32), it is possible to define, for the block depicted in figure 1.32, a high impedance (voltage) input terminal (1), a low impedance (current) input terminal (2), a high impedance (current) output terminal (3) and a low impedance (voltage) output terminal (4). This solution presents all the possible combinations of low and high impedances for both input and output terminals, so this is really a generic block. These features cannot be obtained with voltage or current amplifiers, so the topology chosen as basic active block for LV LP applications is the transconductance amplifier in a transresistance feedback configuration.

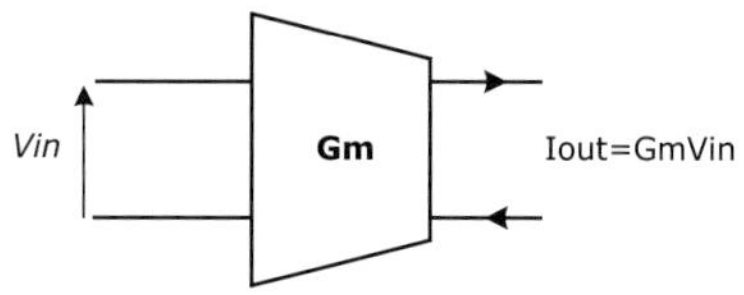

Node	Impedance level
Input	High (ideally ∞)
Output	High (ideally ∞)

Figure 1.31 – Transconductance amplifier

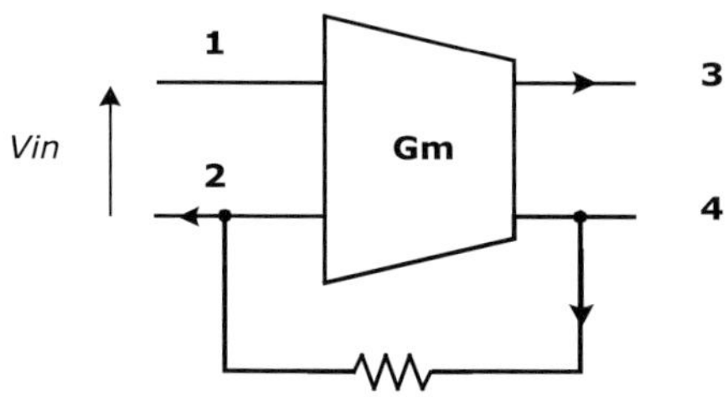

Figure 1.32 – Transconductance amplifier in transresistance feedback configuration

In order to avoid the effect of the output load, a voltage buffer can be inserted, implementing the circuit shown in figure 1.33. This circuit is the so-called *Current Feedback Operational Amplifier* (CFOA) [10,11,12,14,29], whose block scheme and matrix characteristics are reported in figure 1.34.

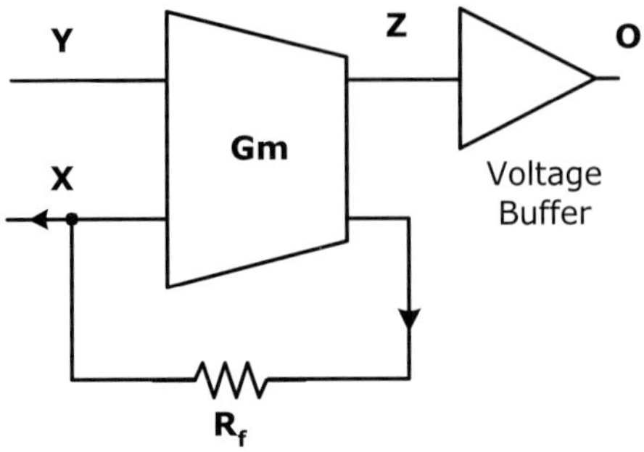

Figure 1.33 – Transconductance amplifier in transresistance feedback with additional output voltage buffer

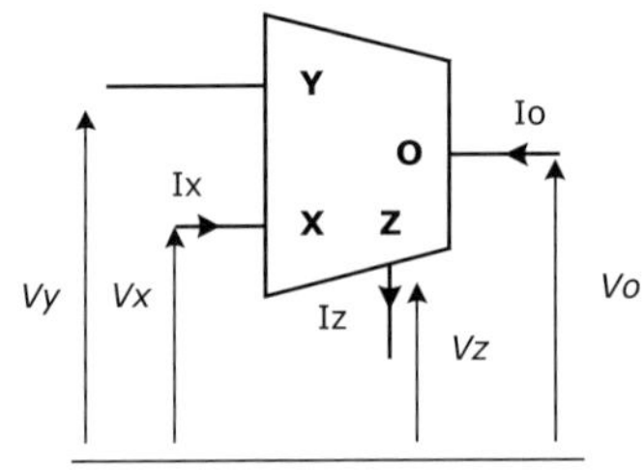

$$\begin{bmatrix} Vx \\ Iy \\ Iz \\ Vo \end{bmatrix} = \begin{bmatrix} 0 & 1 & 0 & 0 \\ 0 & 0 & 0 & 0 \\ 1 & 0 & 0 & 0 \\ 0 & 0 & 1 & 0 \end{bmatrix} \begin{bmatrix} Ix \\ Vy \\ Vz \\ Io \end{bmatrix}$$

Figure 1.34 – CFOA block scheme and matrix characteristics

If only one current output is considered in the chosen OTA, it is possible to derive the CFOA, by applying the current sensing technique and adding a voltage buffer, as reported in figure 1.35.

It has to be noted that, without considering the voltage buffer, the circuit so obtained is a CCII, which will be shown in detail in the next paragraph.

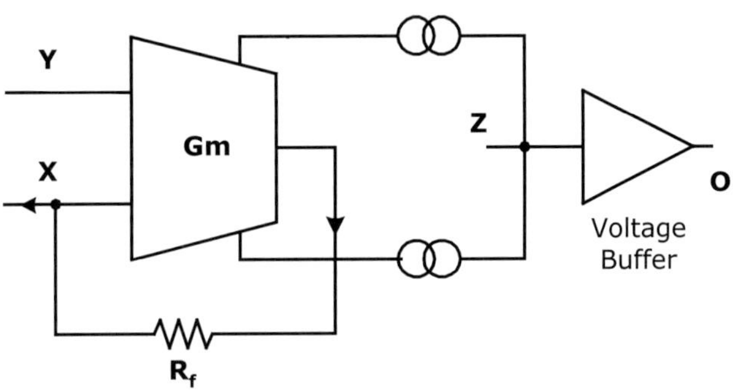

Figure 1.35 – CFOA obtained from a transconductance amplifier in transresistance feedback with current sensing and additional output buffer

Based on the same approach towards "hybrid" amplifiers – which are amplifiers with input and output terminals with different impedance levels referred to different (current and voltage) electrical signals – in the literature [13,30,31] another building block, named *Operational Floating Conveyor* (OFC), has been presented. It is a four terminals active device whose block scheme and matrix characteristics are shown in figure 1.36.

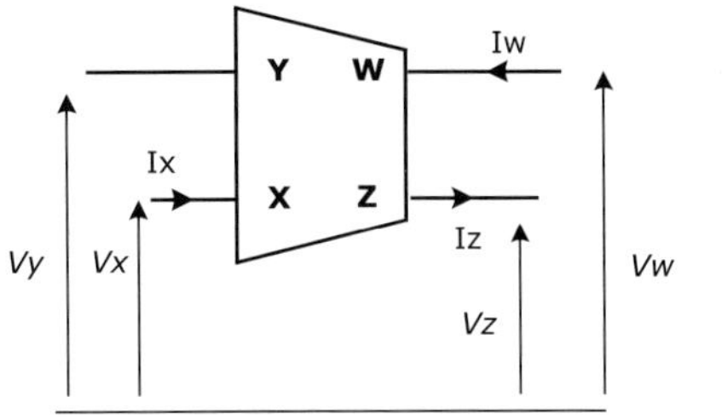

$$\begin{bmatrix} Vx \\ Iy \\ Iz \\ Vw \end{bmatrix} = \begin{bmatrix} 0 & 1 & 0 & 0 \\ 0 & 0 & 0 & 0 \\ 0 & 0 & 0 & -1 \\ Z & 0 & 0 & 0 \end{bmatrix} \begin{bmatrix} Ix \\ Vy \\ Vz \\ Iw \end{bmatrix}$$

Figure 1.36 – OFC block scheme and matrix characteristics

The OFC matrix description is very similar to that of CFOA. In fact, if in the CFOA scheme of figure 1.33, a current sensing is added to the output buffer (see figure 1.37), the further output Z node allows to obtain an OFC [30,31]. In this scheme, the voltage at W node is given by R_w Ix, so R_w replaces Z in the matrix description of figure 1.36.

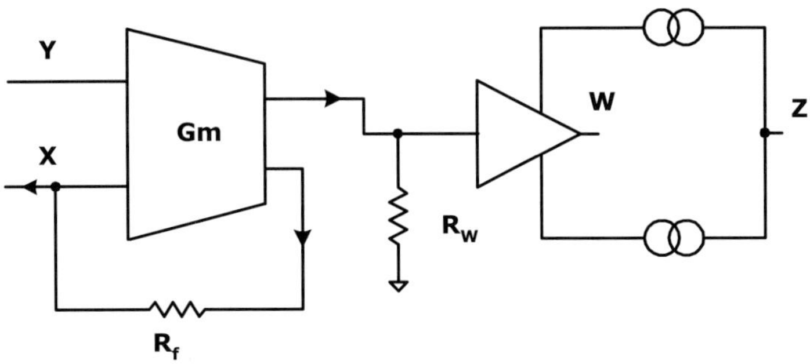

Figure 1.37 – OFC derived from CFOA applying current sensing

Now, since OFC is derived from CFOA by simply applying the supply current sensing principle that utilizes current mirrors at the output terminal, CFOA can be thought to lay inside the OFC, just considering X, Y and W nodes. Thanks to its versatility and utility, OFC can be regarded itself as a universal basic block.

In fact, it can be used as an internal element in CFOA (figure 1.38(a)), but this block can be used *also* to implement a CCII, as shown in figure 1.38(b) [30,31].

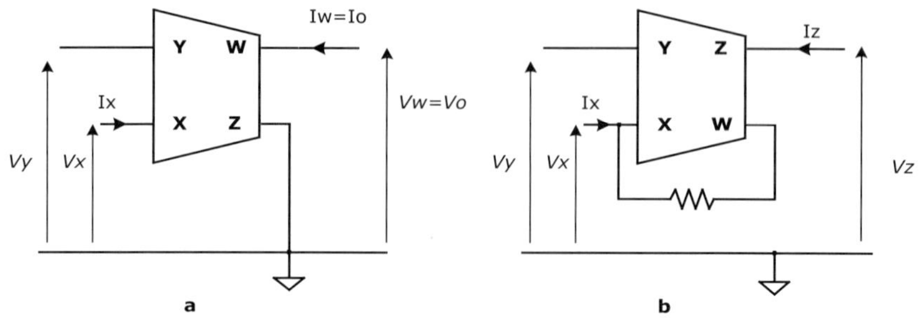

Figure 1.38 – CFOA (a) and CCII (b) derived from OFC

Hence, both OFC and CCII can be seen as natural evolutions of CFOA, the last one being derived from feedback theory.
There is also a different way to see it. From figure 1.33, and more generally from CFOA theory [10,11,12,14], it comes that the same CFOA can be regarded as a CCII followed by a voltage buffer (see figure 1.39).

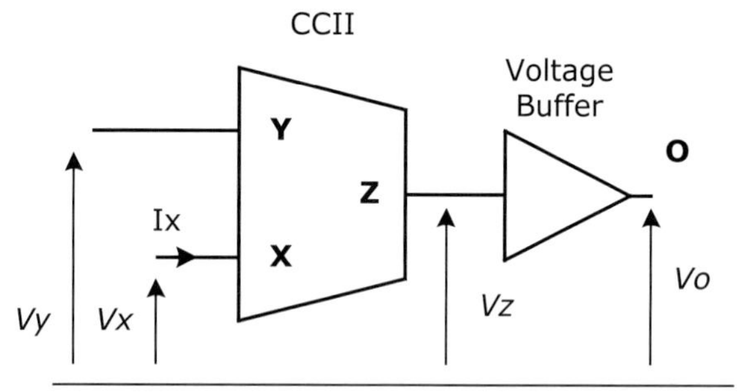

Figure 1.39 – CFOA given by a CCII and a voltage buffer

Basically, CCII is a voltage buffer joined with a current buffer, so CFOA can be also formed by two CCIIs (even if in the second conveyor the output terminal Z is not employed), just replacing the output voltage buffer with the input voltage buffer of a CCII (CCII 2 in figure 1.40).

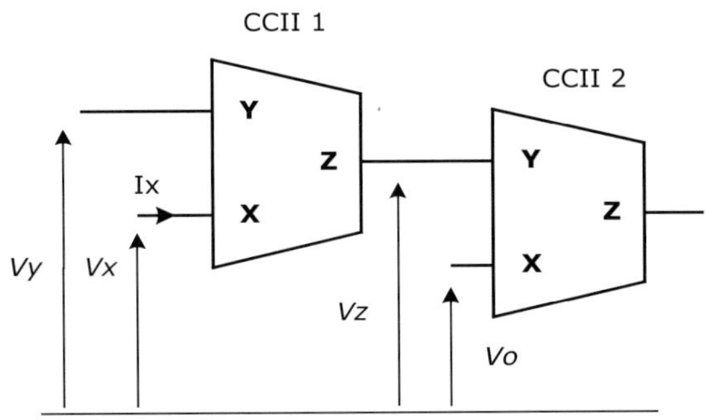

Figure 1.40 – CFOA given by two CCIIs

What happens if the unused current output (Z node of CCII 2) in the circuit, shown in figure 1.40, is taken into account? From the matrix description in fig.1.31, we have an OFC (figure 1.41) implemented through the use of two CCIIs [13].

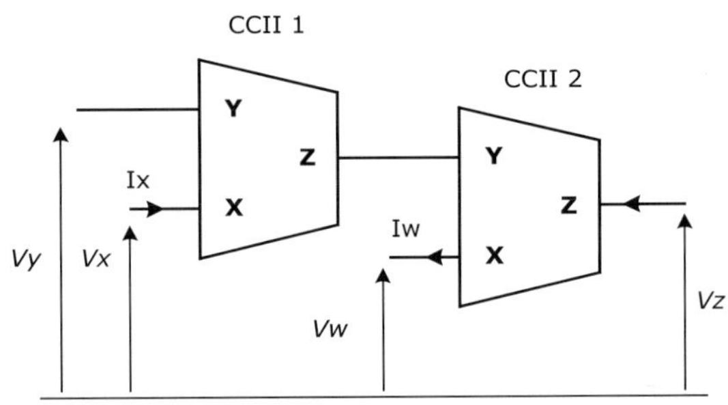

Figure 1.41 – OFC given by two CCIIs

Finally, in the feedback configuration shown in figure 1.32, not considering terminal number 4, it is also possible to identify a high impedance input terminal (number 1), a high impedance output current terminal (2) and a low impedance input/output terminal (3). But this is the CCII!

In conclusion, starting from the more appropriate feedback configuration for LV LP design (the transresistance configuration), different amplifiers, where CCII is always present, have been considered. As a consequence, the new blocks can be simply represented in terms of CCIIs. This is more than a useful schematisation or than a simple application for active elements. This is the reason that allows to consider CCII as the real basic block of analog electronic design [32].

Moreover, among the possible amplifiers, the transconductance type combined with a transresistance feedback configuration allows to obtain a potentially constant bandwidth [8], overcoming the limitation of the constant gain-bandwidth product and having its natural implementation in CCII. As a matter of facts, as all primitive devices – being made up by both a voltage and a current buffer – CCII is really a basic block, because any proposed topology can be regarded as a combination of it, and also can be easily utilised in the implementation of a wide range of analog functions.

1.2.3 A new basic block.

Being a basic building block, CCII can be employed in analog design with a similar equivalent approach used for OAs [33]. From a simplified point of view, the active device behaviour is typically considered as ideal.
For example, a voltage OA is assumed to have infinite voltage gain, infinite input impedance and zero output impedance. Starting from these OA features, an approximate circuit design can be obtained. Obviously, non ideal OA characteristics have to be considered at a deeper design level. Maintaining this design approach, it is possible to implement the same circuits with improved performance using a different basic block. In some cases, this can make the circuit design itself easier. As stated before, CCII can be easily seen as a basic block in analog design, but its success is not comparable to that of OAs. In this paragraph we want to highlight the fact that designers need only a little change in their approach to consider CCIIs at the same level of importance than OA. They have only to deal with the different characteristics of the new block (CCII), while the design flow remains the same.
For example, if a voltage amplification of G is required, the traditional approach utilises an OA with a voltage gain feedback, e.g. the well known feedback configuration reported in figure 1.42 (having negative voltage gain).

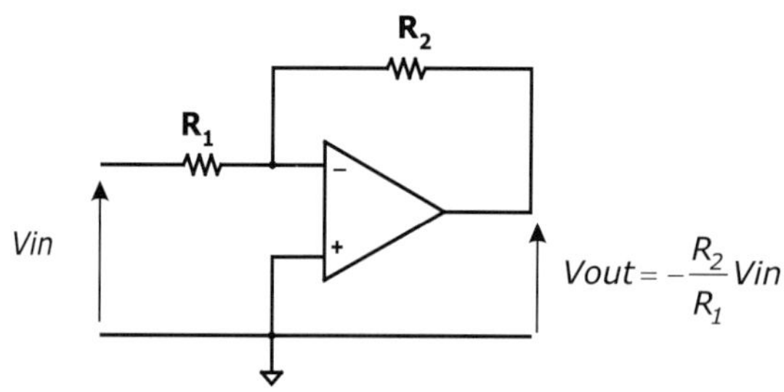

Figure 1.42 – OA-based voltage amplifier

The voltage gain is obtained by choosing R_1 and R_2 values.
Considering CCII as starting block, a voltage amplification can be equivalently obtained with the circuit pictured in figure 1.43.

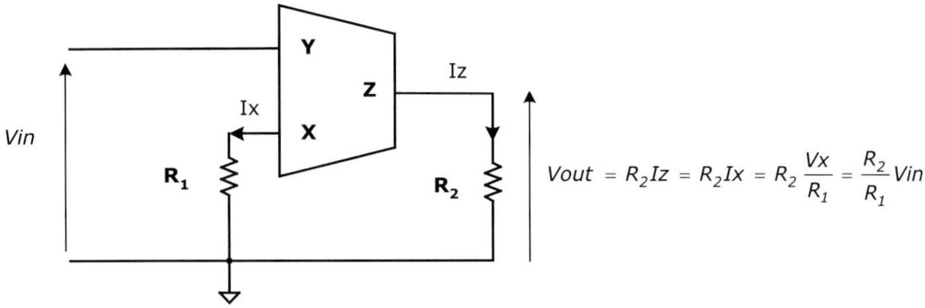

Figure 1.43 – CCII-based voltage amplifier

If a CCII+ is employed, the voltage gain is equal to R_2/R_1; if we use a CCII-, the gain is opposite ($-R_2/R_1$). In the case of OA, two different topologies have to be implemented according to the fact that a negative or a positive gain is needed.
CCII-based voltage amplifier requires the same number of active and passive components than the OA-based one, but CCII implementation does not suffer from the gain-bandwidth limitation of OAs and no current is required at the input.
The comparison between OA-based and CCII-based voltage amplifiers, here shown, is only a simple example. CCII can replace OA in a wide range of applications, often showing better performance. We would like to point out that this brief example has the only aim to show that OA is not always the best choice. Of course, in several applications its use is recommendable, but often the current-mode approach — and particularly CCII — offers an alternative to be considered. As mentioned in the previous sections, other basic blocks such as CFOA and OFC have been proposed in the literature. Each of these active devices has its own peculiarity, so it could be intelligent to use a design approach that chooses the suitable block depending on the specific application. However, from all the considerations done so far, if a basic and general purpose active device is needed, in our opinion, this is the CCII one.

References.

[1] R. Hogervorst, J. H. Huijsing. *Design of low-voltage low-power operational amplifier cell.* Boston: Kluwer Academic Publishers, 1996.

[2] W. A. Serdijin, A. C. van der Voerd, A. H. M. van Roermund, J. Davidse. Design principle for low-voltage low-power analog integrated circuits. Analog Integrated Circuits and Signal Processing. nr. 8; 1998; pp. 115-120.

[3] W. A. Serdijin. *Low-voltage low-power analog integrated circuits*. Boston: Kluwer Academic Publishers, 1995.

[4] C. Toumazou, A. Payne, D. Haigh. *Analogue IC design: The current mode approach*. Peter Peregrinus 1990.

[5] C. Toumazou, J. Lidgey. "Universal current mode analogue amplifiers". In *Analogue IC design: The current mode approach*. Peter Peregrinus 1990.

[6] G. Palumbo, S. Palmisano, S. Pennisi. *CMOS current amplifiers*. Boston: Kluwer Academic Publishers, 1999.

[7] K.Koli, K.Halonen. *CMOS current amplifiers*, Boston, Kluwer Academic Publishers, 2002.

[8] B. Wilson. Transconductance feedback amplifier exhibiting bandwidth independent voltage gain. IEE Proceedings - Circuits Devices and Systems. nr. 5; 1998; pp. 242-248.

[9] B. Wilson. Constant bandwidth amplification using current conveyors. International Journal of Electronics. nr. 65; 1988; pp. 893-898.

[10] S. Franco. Analytical foundation of current feedback amplifiers. Proceedings of the IEEE International Symposium on Circuits and Systems. 1993; Chicago (USA).

[11] D. F. Bowers. "Applying current feedback to voltage amplifiers". In *Analogue IC design: The current mode approach.* Peter Peregrinus 1990.

[12] A. Soliman. Applications of the current feedback operational amplifier. Analog Integrated Circuits and Signal Processing. nr. 11; 1996; pp. 265-302

[13] C. Toumazou, A. Payne, J. Lidgey. Current-feedback versus voltage amplifiers: Hystory, Insight and Relationships. Proceedings of the IEEE International Symposium on Circuits and Systems. 1993. Chicago (USA).

[14] G. Palumbo and S. Pennisi. Current feedback amplifiers versus voltage operational amplifiers. IEEE Transactions on Circuit and Systems-I. nr. 5; 2001; pp. 617-623.

[15] A. Sedra, K. C. Smith. The current conveyor – A new circuit building basic block. IEEE Proceedings. nr. 56; 1968; pp. 1368-1369.

[16] A. S. Sedra, G. W. Roberts. "Current conveyor theory and practice". In *Analogue IC design: The current mode approach.* Peter Peregrinus 1990.

[17] A. Sedra, K. C. Smith. A second generation current conveyor and its applications. IEEE Transactions on Circuit Theory. CT-17; 1970; pp. 132-134.

[18] A. Fabre, H. Amrani, H. Barthelemy. Novel class AB first generation current conveyor. IEEE Transactions on Circuit and Systems-II. nr. 1; 1999; pp. 96-98.

[19] E. Brunn. Class AB CMOS first generation current conveyor. Electronics Letters. nr. 6; 1995; pp. 422-423.

[20] R. Cabeza, A. Carlosena. Analog universal active Device: Theory, design and applications. Analog Integrated Circuits and Signal Processing. nr. 12; 1997; pp. 153-168.

[21] B. H. Tellegen. La recherche pour une serie complete d'elements de circuit ideaux non-lineaires. Rendiconti del seminario matematico e fisico di Milano 1954; Milano.

[22] H. J. Carlin. Singular network elements. IEEE Transactions on Circuit Theory. CT-11; 1964; pp. 67-72.

[23] C. Toumazou, A. Payne. Analog amplifiers: classification and generalization. IEEE Transactions on Circuit and Systems-I. nr. 1; 1996; pp. 43-50.

[24] E. M. Cherry. Feedback amplifier configurations. IEE Proceedings - Circuits Devices and Systems. nr. 6; 2000; pp. 334-346.

[25] A. Arbel. Negative Feedback Revisited. Analog Integrated Circuits and Signal Processing. nr. 10; 1996; pp. 157-178.

[26] G. Palumbo, J. Choma jr. An overview of analog feedback, part I: Basic Theory. International Journal of Analog Integrated Circuits and Signal Processing. nr. 3; vol. 17; 1998.

[27] W. A. Serdijin, J. Mulder, D. Rocha, L. C. C. Marques. Advances in low-voltage low power analog circuit design. Proceedings of the IEEE International Conference on Electronic Circuits and Systems, 2001; Malta.

[28] E. Bruun. Feedback analysis of transimpedance operational amplifier circuits. IEEE Transactions on Circuit and Systems-I. nr. 4; 1993; pp. 275-277.

[29] A. Arbel, L. Magran. Current-mode feedback amplifier employing a transistorized feedback network. Proceedings of the IEEE International Symposium on Circuits and Systems, 1994; London.

[30] C. Toumazou, A. Payne, J. Lidgey. Operational Floating Conveyor. Electronics Letters. nr. 8; 1991; pp. 651-652.

[31] C. Toumazou, A. Payne. Operational Floating Conveyor. Proceedings of the IEEE International Symposium on Circuits and Systems, 1991.

[32] J. Lidgey, K. Hayatleh. Are current conveyors finally coming of age? Electronics World. April; 2000; pp. 322-324.

[33] R. Carbeza, A. Carlosena, A. Arbel. Use of a CCII- as a universal building block. Microelectronics Journal. nr. 28; 1997; pp. 543-550.

CHAPTER II

DESIGN OF CCII TOPOLOGIES

In this chapter, some circuit solutions for the implementation of CCIIs in CMOS technology are proposed. The ideal features of a second generation current conveyor will be introduced in detail and some CCII equivalent models will be investigated.
Then, a wide number of topologies for CCII will be described and analysed, so considering the differences between simulated and theoretical results.

2.1 CCII IDEAL AND REAL CHARACTERISTICS AND EQUIVALENT MODELS

2.1.1 The ideal current conveyor.

In the first chapter the basic characteristics of a second generation current conveyor have been introduced. Its block representation is reported again in figure 2.1 [1,2].

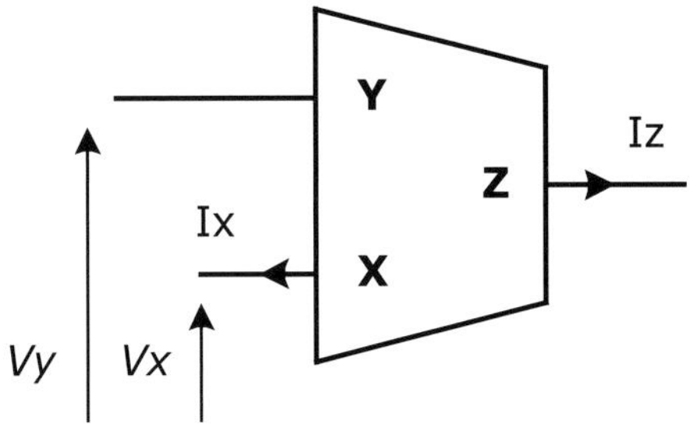

Figure 2.1 - CCII block representation

In a matrix representation, the behaviour of CCII signals, voltages and currents, has been proposed too, so summarising the overall response of the block (figure 2.2).

$$\begin{bmatrix} Iy \\ Vx \\ Iz \end{bmatrix} = \begin{bmatrix} 0 & 0 & 0 \\ 1 & 0 & 0 \\ 0 & \pm 1 & 0 \end{bmatrix} \begin{bmatrix} Vy \\ Ix \\ Vz \end{bmatrix}$$

Figure 2.2 – CCII complete matrix description

The impedance level of the terminals has been also considered. Those reported in figure 2.3 are the ideal ones. The CCII ideal equivalent model is shown in figure 2.4.

CCII Node	Impedance level
X	0
Y	∞
Z	∞

Figure 2.3 – CCII ideal impedance levels

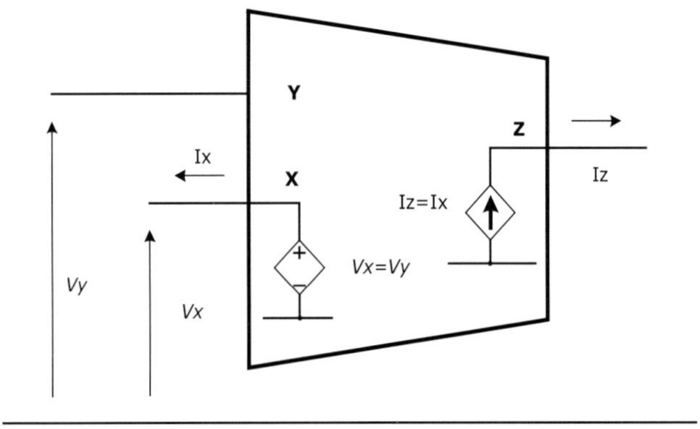

Figure 2.4 –CCII ideal equivalent model

2.1.2 The real current conveyor.

The circuit implementation of CCIIs leads unavoidably to design devices whose characteristics are close, but not equal, to the ideal ones, described in the previous paragraph.
In this section these differences will be considered in detail and a practical model for the second generation current conveyor will be introduced.
Figure 2.5 shows a first model of a real, or non-ideal, CCII. α and β parameters have been introduced to consider the non-perfect voltage and current buffer characteristics, even if real values for α and β are very close to unity.

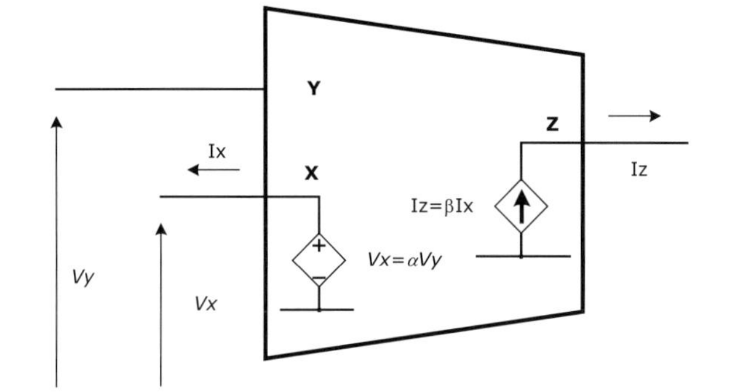

Figure 2.5 – Non ideal CCII model

In figure 2.6 the ideal and non-ideal equivalent models for CCII X node are reported, first considering a non-ideal voltage buffer characteristic and then also a non-zero impedance at the same terminal.

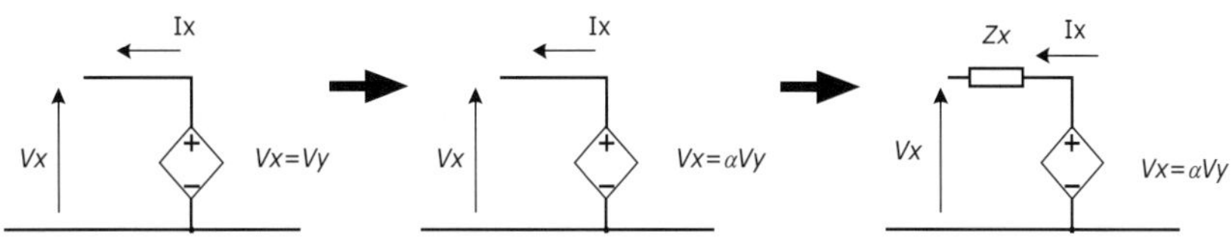

Figure 2.6 –X node equivalent models

Figures 2.7 a and b show the same ideal and real equivalent models for what concerns Z node, in CCII+ and CCII- cases, respectively.

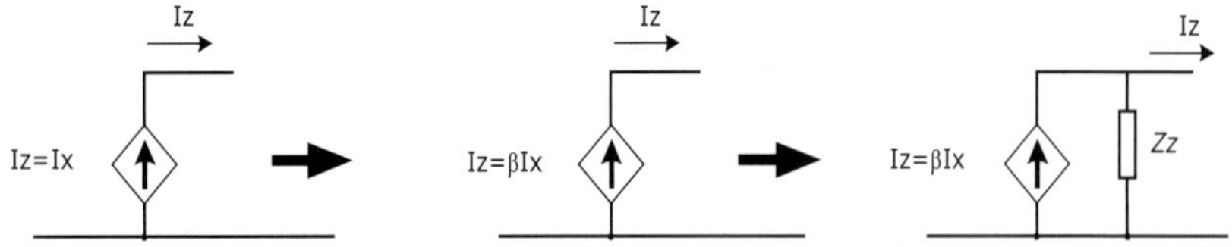

Figure 2.7 a – Z node equivalent models (CCII+)

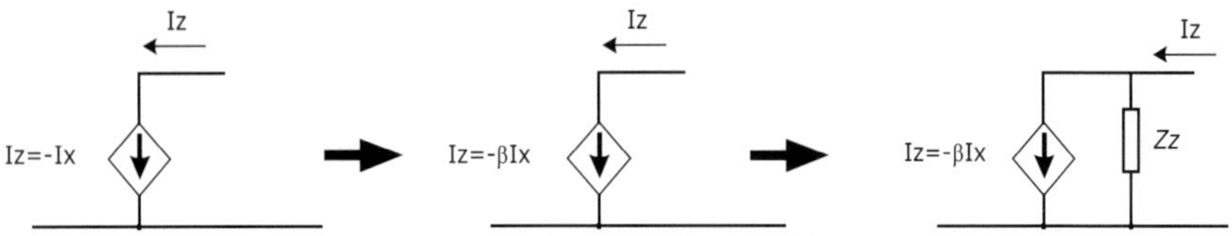

Figure 2.7 b – Z node equivalent models (CCII-)

Hence, figure 2.8 shows the complete real equivalent CCII+ and CCII- models, which take into account the non-ideal characteristics at X and Z nodes, and also the capacitive parasitic impedance at Y node. Some circuit solutions reducing the offset voltage between X and Y nodes are reported in [3] and analysed in a more detail in the next chapter.
Other CCII models and related considerations are reported in [4,5,6,7,8,9,10].

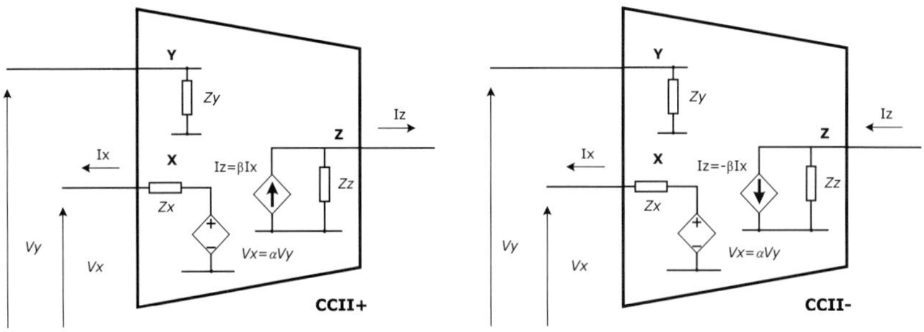

Figure 2.8 – Non-ideal CCII+ and CCII- equivalent models

2.2 CCII TOPOLOGIES: CIRCUIT SCHEMES AND ELECTRICAL CHARACTERISTICS

CCII has been introduced as a "natural" building block in analog circuit design. In other words, it has the same meaning that transistor has at a lower design level. The nMOS transistor can be regarded as a CCII, as shown in figure 2.9. This not only gives the idea of the importance and usefulness of CCII, but also introduces a particular analogy between transistor and conveyor [1].

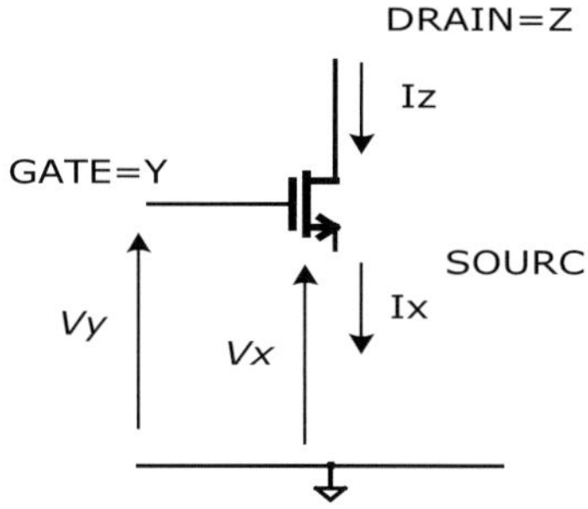

Figure 2.9 - nMOS transistor and its equivalence with CCII

In fact, owing to the source follower effect, the signal applied to Y node (MOS gate) is almost equal to that obtained at X node (source), as expressed by the following α parameter, which represents the node X-Y voltage transfer function:

$$\alpha = \frac{V_X}{V_Y} = \frac{g_m r_0 R_{XLOAD}}{1 + g_m r_0 R_{XLOAD}} \cong 1 \tag{2.1}$$

Ix and Iz currents are equal, as cleared from the small signal equivalent circuit of the nMOS transistor, shown in figure 2.10. The ratio between these currents is expressed by the following β parameter, which is exactly equal to 1 in this analogy:

$$\beta = \frac{I_Z}{I_X} = 1 \tag{2.2}$$

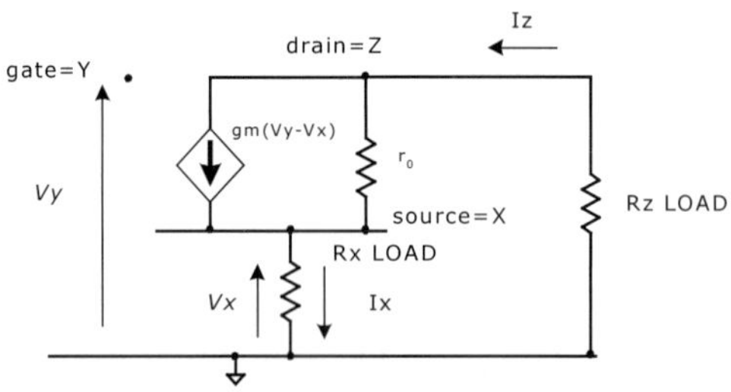

Figure 2.10 – Equivalent circuit for small signal analysis

Impedance level at Y node is given by the transistor gate capacitance, so it is quite high, as required by theoretical specifications. X node impedance value is affected by the load connected to Z node, while the impedance seen at Z terminal is related to the load connected to X node. In particular, being γ a constant parameter whose value is 2/3 in saturation region and 1 otherwise, W and L width and length of the MOS, respectively, and C_{OX} the unitary gate capacitance, we have :

$$Z_Y = \gamma WLC_{OX} \tag{2.3}$$

$$Z_X \cong \frac{r_0 + R_{ZLOAD}}{1 + g_m r_0} \cong \frac{1}{g_m} \quad , \quad \text{if } r_0 >> R_{ZLOAD} \tag{2.4}$$

$$Z_Z = r_0 + (1 + g_m r_0)\, R_{XLOAD} \tag{2.5}$$

A single MOS transistor can be seen as a second generation current conveyor considering small input signals, but biasing voltages at X and Y show a relative difference of about one threshold voltage. This difference can be eliminated considering a traditional nMOS current mirror, which can also be seen as a CCII, as shown in figure 2.11 [1].

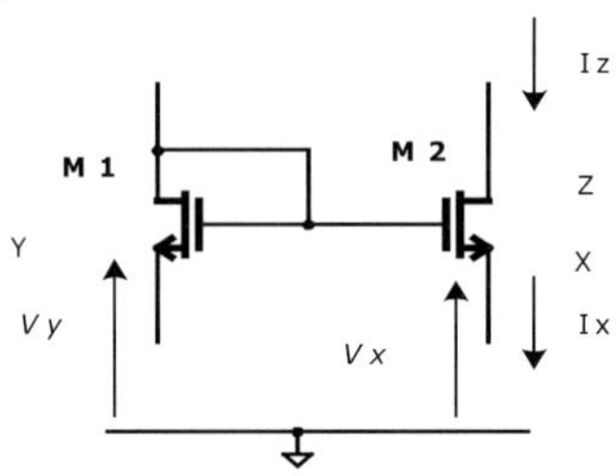

Figure 2.11 - CCII characteristics in the nMOS traditional current mirror

Concerning this circuit, the current transfer function β and the impedance levels at X and Z node are the same presented before (equations (2.2),(2.4),(2.5)). On the contrary, from figure 2.12 it can be determined the following Y node impedance:

$$Z_Y = R_{IBIAS\,2} // \left(\frac{1}{gm_1} + R_{IBIAS\,1} \right)$$

(2.6)

The voltage at X node can be expressed as a function of the voltage at M2 gate (V_{G2}), as follows :

$$\frac{V_X}{V_{G2}} = \frac{g_{m2} r_{02} R_{XLOAD}}{1 + g_{m2} r_{02} R_{XLOAD}} \cong 1$$

(2.7)

In order to evaluate the α parameter, V_{G2} has been determined as a function of V_Y, considering the circuits depicted in figure 2.12. We can write :

$$V_{G2} = \frac{R_{IBIAS\,1}}{R_{IBIAS\,1} + \dfrac{r_{01}}{1 + g_{m1} r_{01}}} V_Y \cong V_Y$$

(2.8)

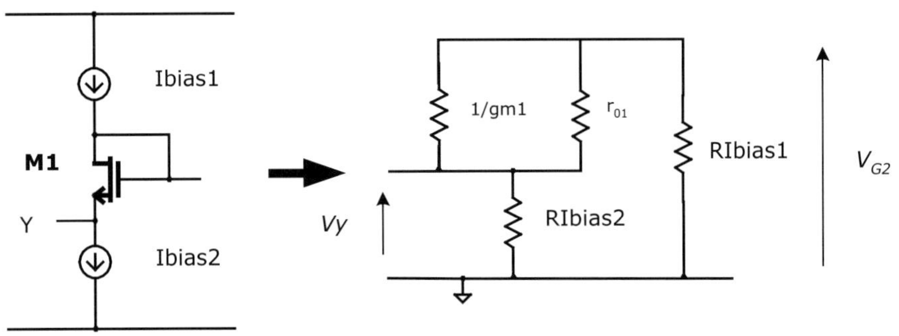

Figure 2.12 - Half nMOS current mirror topology and small signal equivalent circuit

From eq.(2.8) it comes that the voltage V_{G2} is equal to the one applied at Y node only if the biasing sources show a large output resistance. In this way, we obtain the same formula for α parameter (equation (2.1)).

In figure 2.11 a negative CCII has been implemented. If a positive current conveyor is needed, it is possible to add a current mirror, as pictured in figure 2.13. In this topology, Ix and Iz flow in the same direction with respect to the CCII, so performing the CCII+ operation.

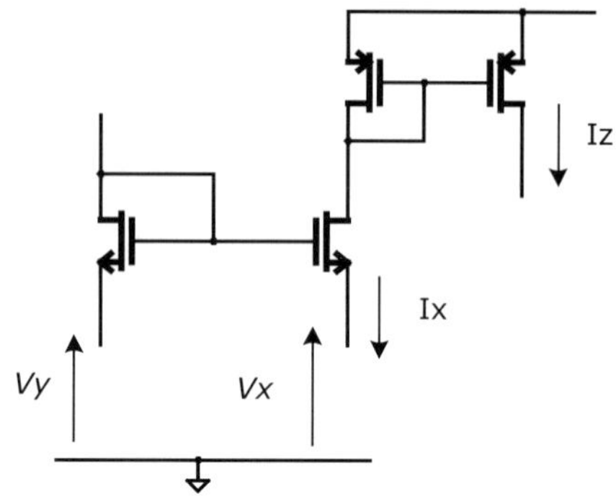

Figure 2.13 - Current mirrors used for positive CCII implementation

The topology presented in figure 2.13 can be "doubled" to obtain a class AB current conveyor, shown in figure 2.14. In this circuit, obviously, I_{Bias1} and I_{Bias2} have to be equal [6,10].

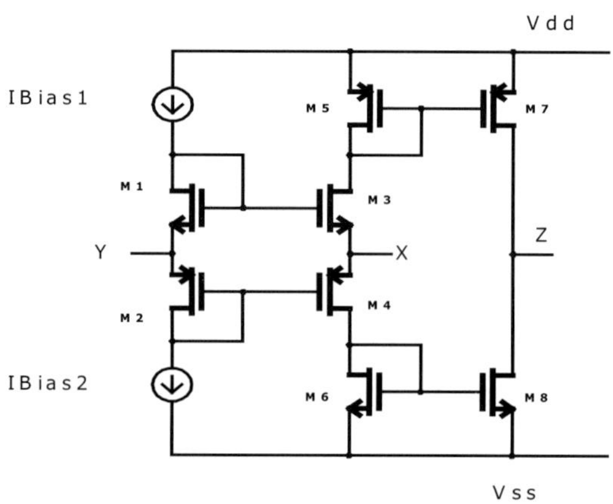

Figure 2.14 - CCII topology based on current mirrors

Considering the products $g_m r_0$ much greater than 1, the voltage characteristic is very close to the ideal one. In fact:

$$\alpha = \frac{V_X}{V_Y} = \frac{1}{1 + \dfrac{1}{(g_{m3} + g_{m4})(r_{03} // r_{04})}} \cong 1 \qquad (2.9)$$

If the loads connected to X and Z nodes show negligible impedances with respect to the output resistance of the MOS transistors, and considering always $g_m r_0 >> 1$, β factor can be simply derived. Its value has been considered for resistive values of the loads connected to CCII terminals lower than the MOS output resistances. A straightforward analysis brings to the following simplified expression for β :

$$\beta = \frac{I_Z}{I_X} \cong \frac{g_{m3}g_{m6}g_{m7} + g_{m4}g_{m5}g_{m8}}{g_{m5}g_{m6}(g_{m3} + g_{m4})} = 1 \quad \text{if} \quad g_{m5} = g_{m7} \quad \text{and} \quad g_{m6} = g_{m8} \qquad (2.10)$$

The impedance level at Y node can be affected by the output impedance of the two current sources, as will be clarified later. A high impedance level can be ensured employing good biasing sources (showing high resistances), as shown in the following expression of Z_Y :

$$Z_Y = \left(\frac{r_{01}}{1 + g_{m1}r_{01}} + R_{IBIAS\,1}\right) // \left(\frac{r_{02}}{1 + g_{m2}r_{02}} + R_{IBIAS\,2}\right) \cong R_{IBIAS\,1} // R_{IBIAS\,2} \qquad (2.11)$$

In table 2.1 it has not been reported a value for such impedance, but a range of possible values. These values have been obtained using different biasing sources. The formula for the X node impedance is quite complicate, but can be simplified neglecting some components. The following derived expression is in a quite good agreement with the simulated results, as shown also in table 2.1.

$$Z_X \cong \frac{1}{g_{m3} + g_{m4} + \dfrac{r_{03} + r_{04}}{r_{03}r_{04}}} \cong \frac{1}{g_{m3} + g_{m4}} \qquad (2.12)$$

The impedance seen at Z node is typically high and given by:

$$Z_Z = \frac{r_{07} r_{08}}{r_{07} + r_{08}} \tag{2.13}$$

In table 2.1 the main characteristics of the current mirror-based CCII, shown in figure 2.14, are summarised. Theoretical and simulated values of these characteristics are given, showing an acceptable agreement.

Current Conveyor Characteristics				
Data	Value	**Parameter**	Theoretical Value	Simulated Value
Voltage Supply	±1.5 V	**Voltage Gain (α)**	0.998	0.99
Power Consumption	320 μW	**Current Gain (β)**	1.04	1.009 (Rx=Rz=10KΩ)
3dB Bandwidth	445 MHz	**Node Y Parasitic Impedance**	(see text)	from 175.8 KΩ to ∞
Dynamic Range	-400mV, +400mV	**Node X Parasitic Impedance**	1.05 KΩ	0.96 KΩ
Biasing Current IBIAS	26 μA	**Node Z Parasitic Impedance**	302 KΩ	296.2 KΩ - 0.052 pF

Table 2.1 – CCII characteristics for the circuit shown in figure 2.14

The presented solution of CCII has the main advantage of a wide bandwidth, obtained with relatively low biasing currents. Its drawback is represented by the limited dynamic range and by the fact that this is not a very low voltage solution.

2.2.1 Differential pair based CCII.

In literature, a wide number of current conveyors employing a differential pair - which implements, through a feedback connection, the input voltage buffer - has been proposed [11].
In figure 2.15 a first possible solution is presented. The circuit operates in class A and is topologically very simple. A feedback connection at X node provides the low impedance terminal.

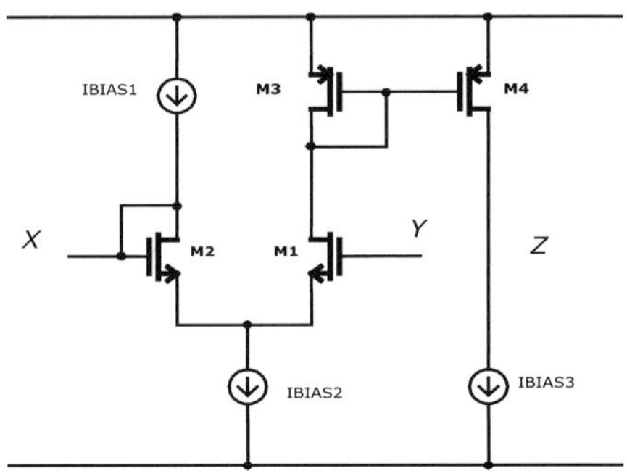

Figure 2.15 - CCII topology based on differential pair

The unity voltage transfer function V_X/V_Y is ensured by the differential pair M1-M2. Large transistors output resistances are required in order to obtain an α parameter dependent only on the input pair transconductance ratio. In fact:

$$\alpha = \frac{V_X}{V_Y} \cong \frac{r_{01} g_{m1}}{1 + r_{02} g_{m2}} \cong \frac{g_{m1}}{g_{m2}} \tag{2.14}$$

Unfortunately, the current transfer function is not independent from the load connected to Z node. If this load impedance is negligible with respect to the transistor output resistances, β parameter is very close to its ideal unitary value, as shown in the following expression:

$$\beta = \frac{I_Z}{I_X} = \frac{r_{03} r_{04} g_{m4}}{(1 + g_{m3} r_{03})(r_{04} + R_{ZLOAD})} \cong \frac{r_{04} g_{m4}}{g_{m3}(r_{04} + R_{ZLOAD})} \cong \frac{g_{m4}}{g_{m3}} \tag{2.15}$$

As seen for the previous topology, the parasitic impedance at Y node is given only by the input transistor gate. Its value is easily evaluated knowing transistor sizes and the unitary capacitance of the input gate.

$$Z_Y = \gamma W(M1) L(M1) C_{OX} \tag{2.16}$$

The parasitic impedance at X node is low, as required, being inversely proportional to the input transistor g_m. A small input resistance requires large values of g_m, so a trade-off between X node parasitic resistance and power consumption has to be made. The value of Z_X has been evaluated considering ideal current sources (see figure 2.15).

$$Z_X = \frac{r_{01} + \frac{r_{02}}{1 + g_{m2} r_{02}}(1 + g_{m1} r_{01}) + \frac{r_{03}}{1 + g_{m3} r_{03}}}{1 + g_{m1} r_{01}} \cong \frac{r_{01} + \frac{g_{m1} r_{01}}{g_{m2}} + \frac{1}{g_{m3}}}{g_{m1} r_{01}} \cong \frac{r_{01}\left(1 + \frac{g_{m1}}{g_{m2}}\right)}{g_{m1} r_{01}} \cong \frac{1}{g_{m1}} + \frac{1}{g_{m2}} \tag{2.17}$$

A high output impedance at Z node is ensured, being Z a drain terminal.

$$Z_Z = \frac{r_{04} R_{IBIAS3}}{r_{04} + R_{IBIAS3}} \tag{2.18}$$

Table 2.2 shows the main characteristics of this current conveyor.

Current Conveyor Characteristics				
Data	**Value**	**Parameter**	**Theoretical Value**	**Simulated Value**
Voltage Supply	±0.75 V	**Voltage Gain (α)**	0.998	0.968
Power Consumption	24 μW	**Current Gain (β)**	1.03	1.06 (Rx=Rz=10KΩ)
3dB Bandwidth	53 MHz	**Node Y Parasitic Impedance**	0.2 pF	0.2 pF
Dynamic Range	0mV, +400mV	**Node X Parasitic Impedance**	14.7 KΩ	14.7 KΩ
Biasing Current IBIAS1	10 μA	**Node Z Parasitic Impedance**	4.69 MΩ	4.44 MΩ

Table 2.2 – CCII characteristics for the circuit shown in figure 2.15

In order to improve the current conveyor performance, the feedback connection at X node can be performed by a nMOS transistor, as pictured in figure 2.16 [11,12].

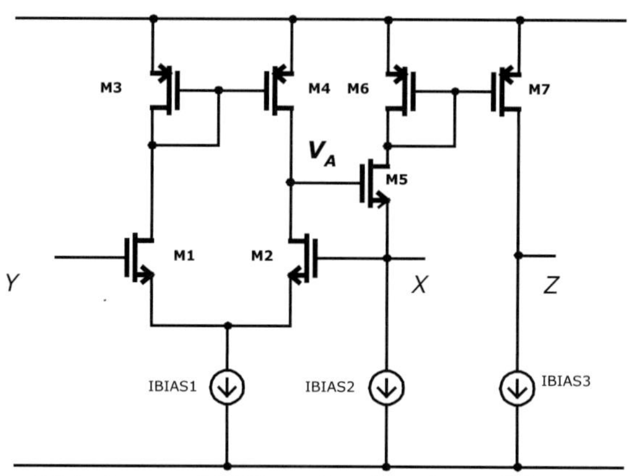

Figure 2.16 – Improved CCII topology based on a differential pair

In this circuit, current mirror M6-M7 is used to sense the current flowing from X node and to mirror it to the high impedance Z node. Also this solution operates in class A.

The determination of α and β parameters can be performed in two different steps. Firstly, we evaluate the expression for V_A (see figure 2.16), in the open loop configuration. Later, this result will be applied to the complete circuit. We have:

$$V_A \cong \frac{r_0}{2}\left(g_{m1}V_Y - g_{m2}V_X\right) \quad \text{where} \quad r_0 \cong r_{01} \cong r_{02} \cong r_{03} \cong r_{04} \tag{2.19}$$

This approximated formula has been derived assuming, as stated above, that all the transistors M1-M4 have the same r_0 value.

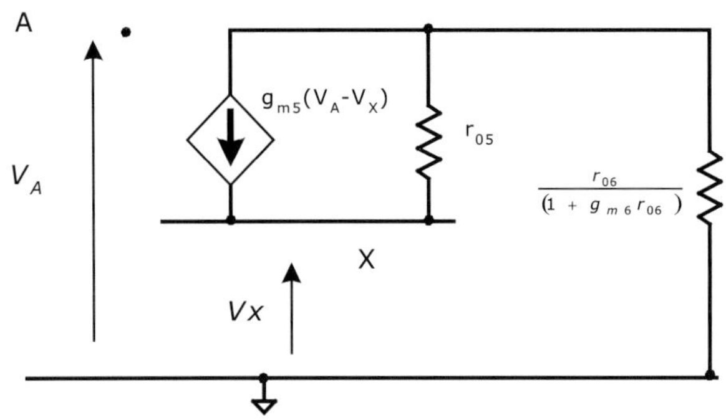

Figure 2.17 – Equivalent circuit for small signal analysis

From the analysis of the circuit in figure 2.17 and using equation (2.19), the α parameter is easily derived. We have:

$$V_X = r_{05}g_{m5}\left(V_A - V_X\right) = r_{05}g_{m5}\left(\frac{r_0}{2}\left(g_{m1}V_Y - g_{m2}V_X\right) - V_X\right) \tag{2.20}$$

from which we can write :

$$\alpha = \frac{V_X}{V_Y} = \frac{r_{05}g_{m5}\frac{r_0}{2}g_{m1}}{1 + r_{05}g_{m5}\left(1 + \frac{r_0}{2}g_{m2}\right)} \cong \frac{g_{m1}}{g_{m2}} \tag{2.21}$$

Once more, α parameter is given by the input transistor transconductance ratio, so it is very close to the ideal unitary value.

β parameter has the same expression of the previous topology. Its value is the following, neglecting the equivalent resistances of the biasing currents I_{BIAS2} and I_{BIAS3} :

$$\beta = \frac{I_Z}{I_X} = \frac{r_{06}r_{07}g_{m7}}{\left(1 + g_{m6}r_{06}\right)\left(r_{07} + R_{ZLOAD}\right)} \cong \frac{r_{07}g_{m7}}{g_{m6}\left(r_{07} + R_{ZLOAD}\right)} \cong \frac{g_{m7}}{g_{m6}} \tag{2.22}$$

Parasitic impedance at Y node is expressed again by equation (2.16), while impedance at X node has been evaluated grounding Y node, in the determined expression for V_A potential (eq. (2.19)), so we have :

$$Z_X = \frac{r_{05} + \dfrac{r_{06}}{1 + g_{m6} r_{06}}}{1 + r_{05} g_{m5}\left(1 + \dfrac{r_0}{2} g_{m2}\right)} \cong \frac{r_{05} + \dfrac{1}{g_{m6}}}{r_{05} g_{m5}\left(1 + \dfrac{r_0}{2} g_{m2}\right)} \cong \frac{1}{g_{m5}\left(1 + \dfrac{r_0}{2} g_{m2}\right)} \cong \frac{2}{g_{m5} r_0 g_{m2}}$$

(2.23)

From eq.(2.20) we can see that if V_Y=0, V_A voltage is:

$$V_A = -g_{m2} \frac{r_0}{2} V_X$$

(2.24)

If we consider the frequency response of the differential pair (see figure 2.16), we can write:

$$V_A = -\frac{g_{m2} \dfrac{r_0}{2}}{1 + \dfrac{s}{p_0}} V_X$$ where p_o is the dominant pole

(2.25)

Using eq.(2.25) for the X node impedance evaluation, we can write:

$$Z_X \cong \frac{2}{g_{m5} r_0 g_{m2}}\left(1 + \frac{s}{p_0}\right) = \frac{2}{g_{m5} r_0 g_{m2}} + \frac{2 \dfrac{s}{p_0}}{g_{m5} r_0 g_{m2}}$$

(2.26)

It is clear that the parasitic impedance at X node shows an inductive component whose value is related to the resistive part and to the dominant pole of the differential pair frequency response. The same approach which considers a dominant pole has been followed for all the other topologies presented in this chapter.

The parasitic impedance obtained at X node is much lower than the one derived for the previous topology. Even if the g_m value is not very high, a really low impedance can be obtained, reduced by the factor $r_0 g_{m5}$ with respect to the value reported in eq.(2.17). This is a good advantage for this topology.

Z node impedance is always high, as required by current conveyor theoretical characteristics, and given by :

$$Z_Z = \frac{r_{07}R_{IBIAS3}}{r_{07} + R_{IBIAS3}} \tag{2.27}$$

Table 2.3 shows the main CCII characteristics of the circuit shown in figure 2.16.

Current Conveyor Characteristics					
Data	**Value**	**Parameter**	**Theoretical Value**	**Simulated Value**
Voltage Supply	± 1 V	**Voltage Gain (α)**	0.999	0.996
Power Consumption	58 μW	**Current Gain (β)**	1.019	1.015 (Rx=Rz=10KΩ)
3dB Bandwidth	19 MHz	**Node Y Parasitic Impedance**	0.4 pF	0.4 pF
Dynamic Range	-200mV, +200mV	**Node X Parasitic Resistance**	27 Ω	25.5 Ω
Biasing Current IBIAS1	10 μA	**Node X Parasitic Inductance**	114 μH	90 μH
Biasing Current IBIAS2=IBIAS3	5 μA	**Node Z Parasitic Resistance**	2.75 MΩ	2.45 MΩ

Table 2.3 – CCII characteristics for the circuit shown in figure 2.16

The previous topology is based on the feedback applied to differential pair through the use of a transistor (M5). The current flowing from X node is sensed through M6 and mirrored to the high impedance Z terminal.

A different circuit solution that uses the same concept is presented in figure 2.18. This circuit shows improved characteristics at lower supply voltages [12,13,14].

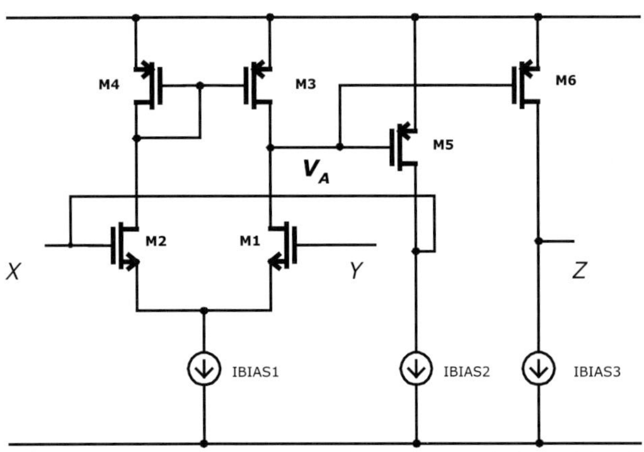

Figure 2.18 – Simplified CCII topology based on differential pair

The output voltage at X node is taken at M5 transistor drain. This voltage is once more controlled by V_A and directly connected to X node to implement the feedback which guarantees a low impedance at X node. In this case V_A is given by:

$$V_A \cong \frac{r_0}{2}\left(g_{m2}V_X - g_{m1}V_Y\right) \quad \text{where} \quad r_0 \cong r_{01}//r_{03} \tag{2.28}$$

A straightforward circuit analysis brings to the following expression for α parameter (eq.(2.29)), which is very similar to that derived for the previous topology (eq.(2.21)).

$$\alpha = \frac{V_X}{V_Y} = \frac{r_{05}\, g_{m5} \frac{r_0}{2} g_{m1}}{r_{05}\, g_{m5} \frac{r_0}{2} g_{m2} + 1} \cong \frac{g_{m1}}{g_{m2}} \quad \text{where} \quad \frac{r_0}{2} = r_{01}//r_{03} \tag{2.29}$$

A drawback of this solution may be represented by the dependence of β on the two loads connected to X and Z nodes - whose resistive values have to be lower of that of typical MOS output resistances - while in the previous topology only the load connected to Z node could affect β (see eq.(2.22)). In this case, if $r_{05} >> R_{XLOAD}$ and $r_{06} >> R_{ZLOAD}$, we have:

$$\beta = \frac{I_Z}{I_X} = \frac{\dfrac{r_{06}}{r_{06} + R_{ZLOAD}} g_{m6}}{\dfrac{r_{05}}{r_{05} + R_{XLOAD}} g_{m5}} \cong \frac{g_{m6}}{g_{m5}} \tag{2.30}$$

For what concerns Z_Y impedance, eq.(2.16) still holds, while X node parasitic impedance has a resistive contribution, given by:

$$R_X = \frac{1}{\dfrac{1}{r_{05}} + g_{m2} \dfrac{r_0}{2} g_{m5}} \cong \frac{2}{g_{m2} r_0 g_{m5}} \tag{2.31}$$

The simplified value of node X impedance is equal to that obtained in the previous topology (eq.(2.23)). Following the same approach shown before, it is possible to demonstrate that the parasitic impedance at X node shows an inductive component too, given by:

$$L_X \cong \frac{\frac{2}{p_0}}{g_{m5} r_0 g_{m2}} \tag{2.32}$$

where p_0 is still the dominant pole of differential pair frequency response. Node Z impedance is typically high and given by:

$$Z_Z = \frac{r_{06} R_{IBIAS3}}{r_{06} + R_{IBIAS3}} \tag{2.33}$$

Table 2.4 shows the main CCII characteristics of the circuit shown in figure 2.18.

Current Conveyor Characteristics				
Data	**Value**	**Parameter**	**Theoretical Value**	**Simulated Value**
Voltage Supply	±0.75 V	**Voltage Gain (α)**	0.992	1.0002
Power Consumption	45 μW	**Current Gain (β)**	1.00	1.00 (Rx=Rz=10KΩ)
3dB Bandwidth	40 MHz	**Node Y Parasitic Impedance**	0.2 pF	0.2 pF
Dynamic Range	0mV, +500mV	**Node X Parasitic Resistance**	27.5 Ω	28 Ω
Biasing Current IBIAS1	10 μA	**Node X Parasitic Inductance**	87 μH	76 μH
Biasing Current IBIAS2= IBAIS3	10 μA	**Node Z Parasitic Impedance**	1.45 MΩ	1.32 MΩ

Table 2.4 – CCII characteristics for the circuit in figure 2.18

The topology introduced in figure 2.18 can be easily modified in order to obtain a class AB current conveyor [13,14,15,16]. This can be done replacing the two biasing sources I_{BIAS2} and I_{BIAS3} with two n-type transistors, as in the circuit presented in figure 2.19.

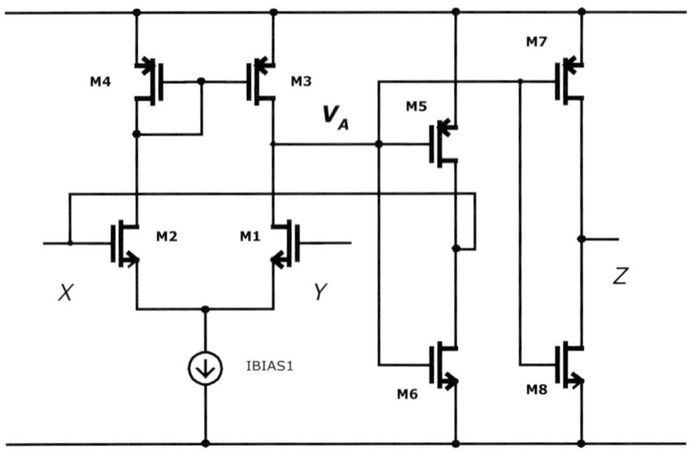

Figure 2.19 – Class AB CCII topology based on differential pair

With respect to the class A CCII, previously introduced, the topology reported in figure 2.19 shows that I_X and I_Z currents can flow both out and in the CCII itself. Moreover, the limitation on the dynamic range is overcome.
The theoretical analysis leads to analytical results very similar to those derived for the previous topology. It is sufficient to replace:

r_{05} with the parallel of r_{05} and r_{06},
r_{06} with the parallel of r_{07} and r_{08},
g_{m5} and g_{m6} with $g_{m5}+g_{m6}$ and $g_{m7}+g_{m8}$, respectively.

If the MOS output resistance is negligible, the voltage transfer error α is only due to the mismatch between the input transistors M1 and M2, so α is always close to 1, as shown in eq.(2.33).

$$\alpha = \frac{V_X}{V_Y} = \frac{\frac{r_{05} r_{06}}{r_{05} + r_{06}} \left(g_{m5} + g_{m6}\right) \frac{r_0}{2} g_{m2}}{\frac{r_{05} r_{06}}{r_{05} + r_{06}} \left(g_{m5} + g_{m6}\right) \frac{r_0}{2} g_{m1} + 1} \cong \frac{g_{m2}}{g_{m1}} \qquad (2.34)$$

For what concerns the current transfer β, the following formula can be easily derived, if the load resistances are not very high when compared to the MOS output resistance.

$$\beta = \frac{I_Z}{I_X} = \frac{\dfrac{\dfrac{r_{07}r_{08}}{r_{07}+r_{08}}}{\dfrac{r_{07}r_{08}}{r_{07}+r_{08}}+R_{ZLOAD}}(g_{m8}+g_{m7})}{\dfrac{\dfrac{r_{05}r_{06}}{r_{05}+r_{06}}}{\dfrac{r_{05}r_{06}}{r_{05}+r_{06}}+R_{XLOAD}}(g_{m6}+g_{m5})} \cong \frac{g_{m8}+g_{m7}}{g_{m6}+g_{m5}}$$

(2.35)

Parasitic impedance at Y node is always capacitive, while X node resistance is easily derived performing the cited substitutions:

$$R_X = \frac{1}{\dfrac{r_{05}+r_{06}}{r_{05}r_{06}} + g_{m2}\dfrac{r_0}{2}(g_{m5}+g_{m6})} \cong \frac{2}{g_{m2}r_0(g_{m5}+g_{m6})}$$

(2.36)

This is valid for the inductive part too, as reported in eq.(2.37).

$$L_X \cong \frac{\dfrac{2}{p_0}}{g_{m2}r_0(g_{m5}+g_{m6})}$$

(2.37)

The Z node output resistance is ensured to be high because it is given by the parallel of two transistor output resistances.

$$Z_Z = \frac{r_{07}r_{08}}{r_{07}+r_{08}}$$

(2.38)

As seen for the previous topologies, the main characteristics and performance of the current conveyor introduced have been summarised in table 2.5.

Current Conveyor Characteristics				
Data	**Value**	**Parameter**	**Theoretical Value**	**Simulated Value**
Voltage Supply	±0.75 V	**Voltage Gain (α)**	1.00	1.0001
Power Consumption	48 μW	**Current Gain (β)**	1.00	1 (Rx=Rz=10KΩ)
3dB Bandwidth	46 MHz	**Node Y Parasitic Impedance**	0.2 pF	0.2 pF
Dynamic Range	–500mV, +500mV	**Node X Parasitic Resistance**	20.3 Ω	20.4 Ω
Biasing Current IBIAS1	10 μA	**Node X Parasitic Inductance**	17.7 μH	18 μH
Current in M5,M6	10 μA	**Node Z Parasitic Impedance**	1.42 MΩ	1.2 MΩ

Table 2.5 – CCII characteristics for the circuit shown in figure 2.19

The use of a differential pair in the implementation of CCIIs can be extended also to the basic topologies. Also the current mirror-based current conveyor, presented in figure 2.14, can be modified employing, as a first stage, a differential pair (see figure 2.20) [17]. Only Y terminal is affected by this modification, while nothing changes for X and Z nodes.

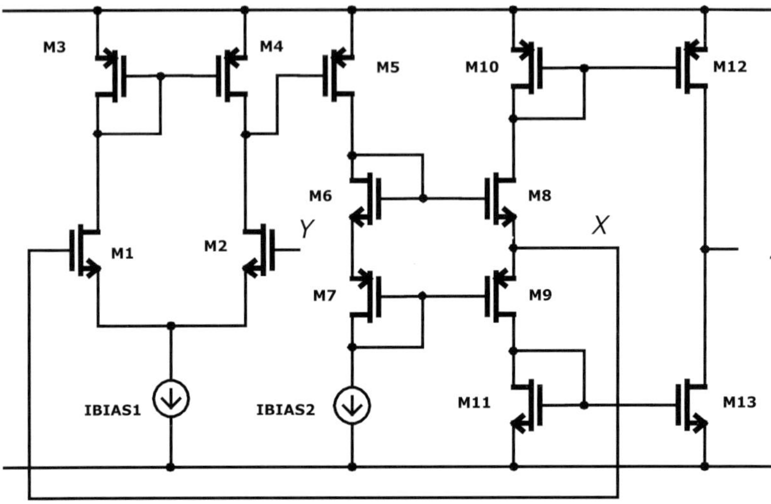

Figure 2.20 – Class AB CCII topology based on differential pair with modified output stage

Using this topology a better response for α parameter can be obtained, as cleared from eq.(2.37).

$$\alpha = \frac{V_X}{V_Y} = \frac{\frac{r_{08} r_{09}}{r_{08} + r_{09}}(g_{m8} + g_{m9})\frac{r_0}{2} g_{m1} r_{05} g_{m5}}{1 + \frac{r_{08} r_{09}}{r_{08} + r_{09}}(g_{m8} + g_{m9}) + \frac{r_{08} r_{09}}{r_{08} + r_{09}}(g_{m8} + g_{m9})\frac{r_0}{2} g_{m2} r_{05} g_{m5}} \cong \frac{g_{m2}}{g_{m1}}$$

(2.39)

A simple circuit analysis brings to the following value for β. The derived expression is very similar to eq.(2.10).

$$\beta = \frac{I_Z}{I_X} \cong \frac{g_{m9} g_{m10} g_{m13} + g_{m8} g_{m11} g_{m12}}{g_{m10} g_{m11}(g_{m8} + g_{m9})} = 1 \quad \text{if} \quad g_{m10} = g_{m12} \ \text{and} \ g_{m11} = g_{m13}$$

(2.40)

The differential pair allows to have a high Y node impedance, independent from the biasing current resistances. This is obviously an improvement with respect to the solution shown in figure 2.14. We have, obviously:

$$Z_Y = \gamma WLC_{OX} \tag{2.41}$$

Thanks to the feedback effect introduced with the differential pair, also the parasitic impedance at X node shows an improvement. In fact, for what concerns the resistive contribution, we have:

$$R_X = \frac{1}{\frac{r_{08} + r_{09}}{r_{08} r_{09}} + \left(1 + g_{m1} \frac{r_0}{2} r_{05} g_{m5}\right)\left(g_{m8} + g_{m9}\right)} \cong \frac{2}{g_{m1} r_0 r_{05} g_{m5} \left(g_{m8} + g_{m9}\right)} \tag{2.42}$$

The only drawback may be represented, with respect to the solution presented in fig. 2.14, by the fact that now the parasitic impedance shows an inductive component, given by:

$$L_X \cong \frac{\frac{2}{p_0}}{g_{m2} r_0 r_{05} g_{m5} \left(g_{m8} + g_{m9}\right)} \tag{2.43}$$

The Z node output impedance remains very high, being given by the transistors output resistances.

$$Z_Z = \frac{r_{012} r_{013}}{r_{012} + r_{013}} \tag{2.44}$$

Table 2.6 summarizes the CCII parameters for the circuit shown in figure 2.20.

Current Conveyor Characteristics				
Data	**Value**	**Parameter**	**Theoretical Value**	**Simulated Value**
Voltage Supply	± 1.5 V	**Voltage Gain (α)**	1.00	1.00
Power Consumption	120 μW	**Current Gain (β)**	1.00	1.00 (Rx=Rz=10KΩ)
3dB Bandwidth	11 MHz	**Node Y Parasitic Impedance**	0.22 pF	0.22 pF
Dynamic Range	-900mV, +500mV	**Node X Parasitic Resistance**	45 mΩ	50 mΩ
Biasing Current IBIAS1	3 μA	**Node X Parasitic Inductance**	45 μH	51 μH
Biasing Current IBIAS2	10 μA	**Node Z Parasitic Impedance**	1.43 MΩ	1.29 MΩ

Table 2.6 – CCII characteristics of the circuit shown in figure 2.20

The implementation of class AB current conveyor allows to improve CCII performance, but a fundamental problem has to be overcome. The biasing currents flowing in the two branches composed by M5 and M6 transistors and by M7 and M8 (see figure 2.19) are controlled only by transistor aspect ratios and depend on the supply voltage. This means that the technological spread could produce biasing currents whose values can result to be very far from those set during the design steps. For this reason a novel solution has been designed, as described in figure 2.21, which implements a different biasing of the output currents [18]. From the DC point of view, the biasing points of M5 and M6 transistors are imposed by suitable choices for I_{BIAS2} currents and R resistance. In this way, the output stages biasing problems are well overcome, because there is no dependence on the supply level.

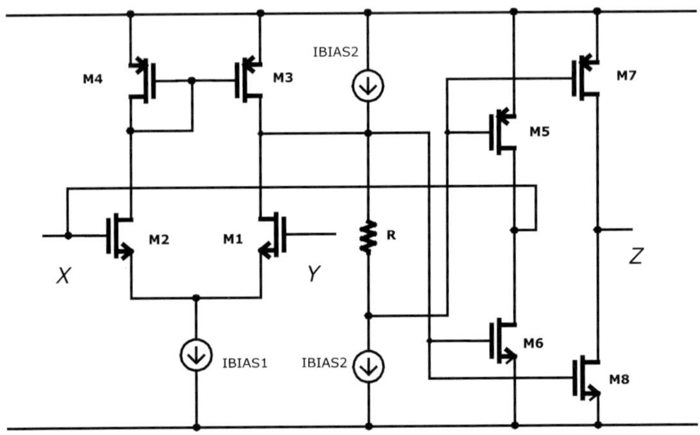

Figure 2.21– Class AB CCII topology with improved biasing solution

Nothing changes if the small signal model is taken into account. In fact if the two biasing sources named I_{BIAS2} in figure 2.21 can be considered ideal, no signal current flows into R resistance, so M5 and M6 gates have the same signal voltage. For this circuit, eq.s (2.30)-(2.33) are still valid.

Current Conveyor Characteristics				
Data	**Value**	**Parameter**	**Theoretical Value**	**Simulated Value**
Voltage Supply	±0.75V	**Voltage Gain (α)**	1.03	0.999
Power Consumption	54 μW	**Current Gain (β)**	1	1 (Rx=Rz=10KΩ)
3dB Bandwidth	27 MHz	**Node Y Parasitic Impedance**	0.2 pF	0.2 pF
Dynamic Range	–200mV, +600mV	**Node X Parasitic Resistance**	27 Ω	29 Ω
Biasing Current IBIAS1	10 μA	**Node X Parasitic Inductance**	27 μH	28 μH
Current in M5,M6	10 μA	**Node Z Parasitic Impedance**	1.43 MΩ	1.26 MΩ

Table 2.7 – CCII characteristics of the circuit shown in figure 2.21

To confirm this statement, simulated results for figure 2.21 circuit have been compared, in table 2.7, to the theoretical ones.
Obviously, it is possible to implement the complementary solution for the input stage, employing a pMOS-based differential pair, as presented in figure 2.22.

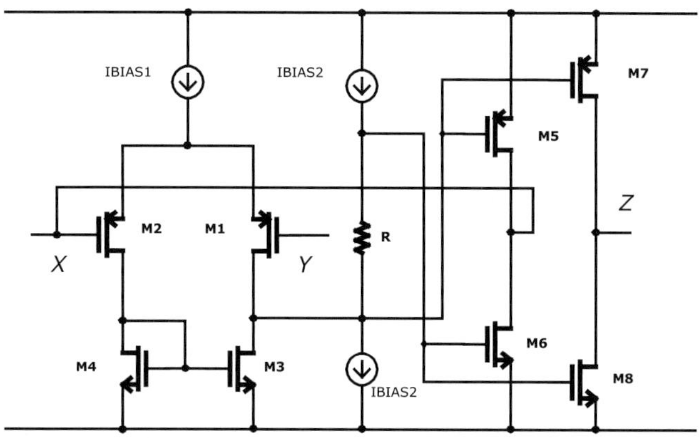

Figure 2.22– Class AB CCII topology with improved biasing solution, based on p-type differential pair

The operating principle is exactly the same seen before. In table 2.8 the main characteristics of the pMOS solution are summarised. From the small signal analysis point of view, formulas (2.30)-(2.33) are always valid. Table 2.8 summarises the CCII parameters for the circuit shown in figure 2.22.

Current Conveyor Characteristics				
Data	**Value**	**Parameter**	**Theoretical Value**	**Simulated Value**
Voltage Supply	±0.75V	**Voltage Gain (α)**	1	0.999
Power Consumption	53 μW	**Current Gain (β)**	1	1 (Rx=Rz=10KΩ)
3dB Bandwidth	28 MHz	**Node Y Parasitic Impedance**	0.2 pF	0.2 pF
Dynamic Range	–750mV, +160mV	**Node X Parasitic Resistance**	30.8 Ω	31.3 Ω
Biasing Current IBIAS1	10 μA	**Node X Parasitic Inductance**	39 μH	35 μH
Current in M5,M6	10 μA	**Node Z Parasitic Impedance**	1.43 MΩ	1.25 MΩ

Table 2.8 – CCII characteristics of the circuit shown in figure 2.22

The last two topologies, depicted in figures 2.21 and 2.22, show a particular behaviour. In fact, in a rough way, we can say that the n-type based CCII operates for high input voltages, while the p-type based operates for low input voltages. This means that the two solutions, if both implemented, seem to be able to manage input signals from the positive (for n-type based) down to the negative (for the p-type based) supply voltage ("rail-to-rail" operation).

The idea is to join the two input stages, placing both the differential pairs "in parallel", so obtaining a "rail-to-rail" current conveyor, as pictured in figure 2.23.

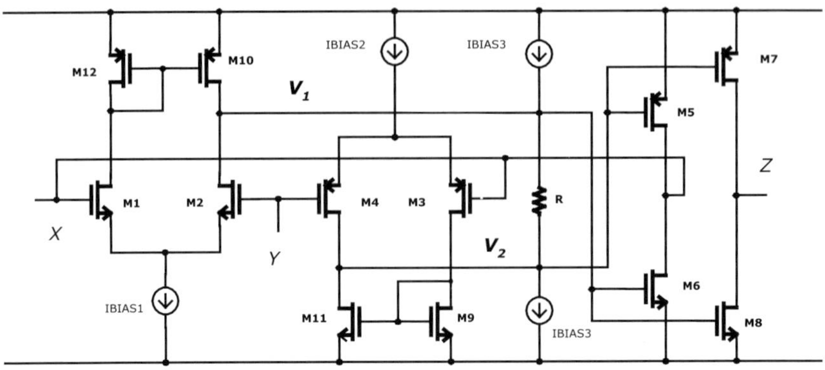

Figure 2.23– Rail-to-Rail Class AB CCII topology with improved biasing solution

When the input voltage applied to Y node is high, the n-type based CCII operates, while the p-type one works for low inputs. In this situation all the theoretical analysis performed in figure 2.21 and 2.22 circuits is still valid. Unfortunately, there will be the input voltage central range for which both the complementary input stages operate. In other words, if one stage (n-type or p-type) is switched off, we can consider (2.30)-(2.33) still valid. If both structures are active, we have to perform a new circuit analysis. The gates of output transistors M5 and M6 (the same is valid for M7 and M8) are biased by two different voltage signals, one from n-type pair (V_1) and another from p-type pair (V_2), as shown in figure 2.23, whose values are the following:

$$V_1 \cong \frac{r_{ON}}{2}(g_{m1}V_X - g_{m2}V_Y) \quad \text{where} \quad \frac{r_{ON}}{2} \cong r_{02}//r_{010}$$
$$V_2 \cong \frac{r_{OP}}{2}(g_{m3}V_X - g_{m4}V_Y) \quad \text{where} \quad \frac{r_{OP}}{2} \cong r_{04}//r_{011} \tag{2.45}$$

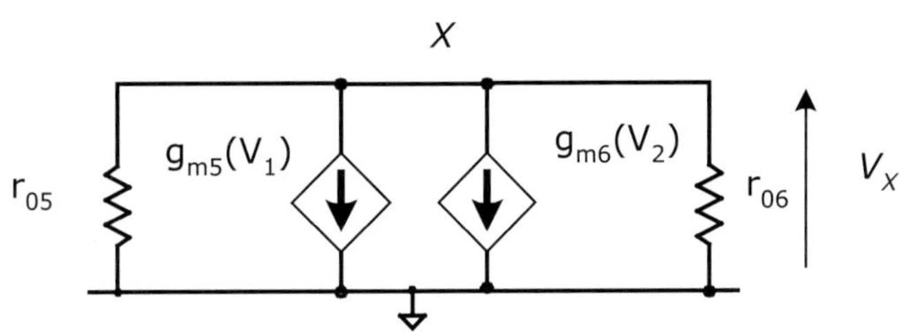

Figure 2.24– Small signal equivalent circuit for the output stage (X node) of the circuit shown in figure 2.23

From figure 2.24 it is possible to write the following expression for V_X:

$$V_X = (g_{m5}V_1 + g_{m6}V_2)\frac{r_{05}r_{06}}{r_{05}+r_{06}}$$

(2.46)

If we substitute eq.(2.45) in eq.(2.46),we determine the following value for α:

$$\alpha = \frac{V_X}{V_Y} = \frac{\frac{r_{05}r_{06}}{r_{05}+r_{06}}\frac{r_{ON}}{2}\frac{r_{OP}}{2}(g_{m5}g_{m2}+g_{m6}g_{m4})}{1+\frac{r_{05}r_{06}}{r_{05}+r_{06}}\frac{r_{ON}}{2}\frac{r_{OP}}{2}(g_{m5}g_{m1}+g_{m6}g_{m3})} \cong \frac{g_{m5}g_{m2}+g_{m6}g_{m4}}{g_{m5}g_{m1}+g_{m6}g_{m3}} = \frac{g_{m2}+g_{m4}\frac{g_{m6}}{g_{m5}}}{g_{m1}+g_{m3}\frac{g_{m6}}{g_{m5}}}$$

(2.47)

From eq.(2.47), it is evident that the real advantage of this circuit is represented by the fact that if the two differential pairs are *separately* well balanced (that is it has to be: $g_{m1}=g_{m2}$ and $g_{m3}=g_{m4}$), α is really close to one. This means that it is not important to have the same transconductance value for both the nMOS and the pMOS input stages.

For what concerns β parameter, a similar situation occurs. In fact, considering eq.s(2.45), (2.46) and the small signal model reported in figure 2.24, the currents that flow into the loads connected to X and Z nodes are:

$$I_X = \frac{\frac{r_{05}r_{06}}{r_{05}+r_{06}}}{\frac{r_{05}r_{06}}{r_{05}+r_{06}}+R_{XLOAD}}(g_{m5}V_1 + g_{m6}V_2)$$

(2.48)

$$I_Z = \frac{\frac{r_{07}r_{08}}{r_{07}+r_{08}}}{\frac{r_{07}r_{08}}{r_{07}+r_{08}}+R_{ZLOAD}}(g_{m7}V_1 + g_{m8}V_2)$$

(2.49)

If R_{XLOAD} and R_{ZLOAD} are negligible with respect to the output resistances r_{0i}, we can derive the following expression for β:

$$\beta = \frac{I_Z}{I_X} = \frac{(g_{m5}V_1 + g_{m6}V_2)}{(g_{m7}V_1 + g_{m8}V_2)} = \frac{g_{m5}}{g_{m7}}\frac{\left(V_1 + \frac{g_{m6}}{g_{m5}}V_2\right)}{\left(V_1 + \frac{g_{m8}}{g_{m7}}V_2\right)} \cong \frac{g_{m5}}{g_{m7}} \cong 1$$

(2.50)

From figure 2.23 it is possible to see that transistors M5, M7 and M6, M8 easily have the same transconductance, while M5, M6 and M7, M8 may also have different g_m values. This ensures a β parameter close to unitary ideal value.
From figure 2.24 small signal scheme, applying an external source in order to evaluate R_X parasitic impedance, we can obtain:

$$R_X = \frac{1}{\frac{r_{05}+r_{06}}{r_{05}r_{06}} + g_{m5}\frac{r_{ON}}{2}g_{m1} + g_{m6}\frac{r_{OP}}{2}g_{m3}} \cong \frac{2}{g_{m5}r_{ON}g_{m1} + g_{m6}r_{OP}g_{m3}} \quad (2.51)$$

If we consider the differential pair having the same small signal parameters ($g_{m1} = g_{m3} = g_m$, $g_{m2} = g_{m4}$, $r_{ON} = r_{OP} = r_0$), eq.(2.51) is equal to eq.(2.36), in fact:

$$R_X \cong \frac{2}{g_{m5}r_{ON}g_{m1} + g_{m6}r_{OP}g_{m3}} \cong \frac{2}{r_0 g_m (g_{m5}+g_{m6})} \quad (2.52)$$

Moreover, an inductive part equal to the one given by eq.(2.37) will result too:

$$L_X \cong \frac{\frac{2}{p_0}}{r_0 g_m (g_{m5}+g_{m6})} \quad (2.53)$$

The expression for Z_Z parasitic impedance is not affected by the modified topology proposed in figure 2.23:

$$Z_Z = \frac{r_{07}r_{08}}{r_{07}+r_{08}} \quad (2.54)$$

Table 2.9 summarizes the CCII parameters for the circuit shown in figure 2.23.

Current Conveyor Characteristics				
Data	**Value**	**Parameter**	**Theoretical Value**	**Simulated Value**
Voltage Supply	±0.75V	**Voltage Gain (α)**	1.005	1.001
Power Consumption	84 μW	**Current Gain (β)**	1.00	1.00 (Rx=Rz=10KΩ)
3dB Bandwidth	25 MHz	**Node Y Parasitic Impedance**	0.4 pF	0.45 pF
Dynamic Range	**-750mV, +640mV**	**Node X Parasitic Resistance**	23 Ω	24.4 Ω
Biasing Current IBIAS1	10 μA	**Node X Parasitic Inductance**	80 μH	70 μH
Current in M5,M6	10 μA	**Node Z Parasitic Impedance**	1.43 MΩ	1.26 MΩ

Table 2.9 – CCII characteristics for the circuit shown in figure 2.23

In the first chapter, second generation current conveyors have been introduced starting from the general feedback theory. It has been shown that they are derived applying a feedback to a transconductance amplifier (OTA). In fact, almost all the presented solutions can be regarded as OTAs with feedback, which confirms once more the theoretical analysis performed in the first chapter. In figure 2.25 a CCII obtained from a simple OTA is reported.

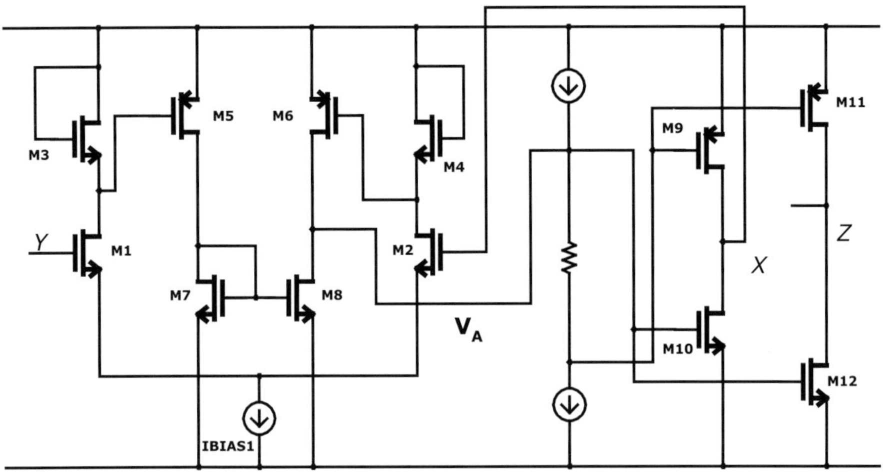

Figure 2.25– Another OTA-based CCII topology

The output voltage of the OTA is connected to an inverter stage, as that presented in figure 2.18. The output of inverter stage is connected back to one input terminal of the OTA, so to have the low impedance X node. Obviously, the output stage can be replaced by its improved version, as in figure 2.21.
Evaluation of α parameter leads to the following expression:

$$\alpha = \frac{V_X}{V_Y} = \frac{\dfrac{r_{08} r_{06}}{r_{08} + r_{06}} \left(\dfrac{g_{m6}}{2 g_{m4}} + \dfrac{g_{m5} g_{m8}}{2 g_{m3} g_{m7}} \right) \dfrac{r_{09} r_{010}}{r_{09} + r_{010}} \left(g_{m9} + g_{m10} \right) g_{m1}}{1 + \dfrac{r_{08} r_{06}}{r_{08} + r_{06}} \left(\dfrac{g_{m6}}{2 g_{m4}} + \dfrac{g_{m5} g_{m8}}{2 g_{m3} g_{m7}} \right) \dfrac{r_{09} r_{010}}{r_{09} + r_{010}} \left(g_{m9} + g_{m10} \right) g_{m2}} \cong \frac{g_{m1}}{g_{m2}} \tag{2.55}$$

The current parameter β is independent from loads connected to X and Z nodes only if their values are negligible with respect to transistor output resistances, as stated in eq.(2.56).

$$\beta = \frac{I_Z}{I_X} = \frac{(g_{m11} + g_{m12})}{(g_{m9} + g_{m10})} \frac{\dfrac{r_{011} r_{012}}{r_{011} + r_{012}}}{\dfrac{r_{011} r_{012}}{r_{011} + r_{012}} + R_{ZLOAD}} \frac{\dfrac{r_{09} r_{010}}{r_{09} + r_{010}} + R_{XLOAD}}{\dfrac{r_{09} r_{010}}{r_{09} + r_{010}}} \cong \frac{g_{m11} + g_{m12}}{g_{m9} + g_{m10}} \tag{2.56}$$

While eq.(2.16) still holds for Z_Y parasitic impedance, Z_X is given by:

$$R_X = \frac{1}{\dfrac{r_{09} + r_{010}}{r_{09} r_{010}} + \left(\dfrac{g_{m6}}{2g_{m4}} + \dfrac{g_{m5} g_{m8}}{2g_{m3} g_{m7}}\right) \dfrac{r_{06} r_{08}}{r_{06} + r_{08}} (g_{m9} + g_{m10}) g_{m2}} \cong \frac{2}{\left(\dfrac{g_{m6}}{g_{m4}} + \dfrac{g_{m5}}{g_{m3}}\right) \dfrac{r_{06} r_{08}}{r_{06} + r_{08}} g_{m2} (g_{m9} + g_{m10})} \tag{2.57}$$

With respect to eq.(2.52) the parasitic impedance at X node is reduced by a factor given by the sum of the ratios of g_{m6}-g_{m4} and g_{m5}-g_{m3}.
Once more an inductive behaviour will result, and the inductive component of the parasitic impedance at X node is:

$$L_X = \frac{1/p_0}{\dfrac{r_{09} + r_{010}}{r_{09} r_{010}} + \left(\dfrac{g_{m6}}{2g_{m4}} + \dfrac{g_{m5} g_{m8}}{2g_{m3} g_{m7}}\right) \dfrac{r_{06} r_{08}}{r_{06} + r_{08}} (g_{m9} + g_{m10}) g_{m2}} \cong \frac{2/p_0}{\left(\dfrac{g_{m6}}{g_{m4}} + \dfrac{g_{m5}}{g_{m3}}\right) \dfrac{r_{06} r_{08}}{r_{06} + r_{08}} g_{m2} (g_{m9} + g_{m10})} \tag{2.58}$$

The output impedance at Z node is high and simply due to the parallel of r_{011} and r_{012}.

$$Z_Z = \frac{r_{011} r_{012}}{r_{011} + r_{012}} \tag{2.59}$$

Table 2.10 summarises the CCII parameters for the circuit shown in figure 2.25.

Current Conveyor Characteristics				
Data	**Value**	**Parameter**	**Theoretical Value**	**Simulated Value**
Voltage Supply	±0.75 V	**Voltage Gain (α)**	0.984	0.999
Power Consumption	80 μW	**Current Gain (β)**	1.00	1 (Rx=Rz=10KΩ)
3dB Bandwidth	28 MHz	**Node Y Parasitic Impedance**	0.09 pF	0.07 pF
Dynamic Range	-350mV - +700mV	**Node X Parasitic Resistance**	61 Ω	55 Ω
Biasing Current IBIAS1	10 μA	**Node X Parasitic Inductance**	78 μH	140 μH
Current in M9,M10	10 μA	**Node Z Parasitic Impedance**	1.43 MΩ	1.29 MΩ

Table 2.10 – CCII characteristics for the circuit shown in figure 2.25

Another solution based on OTA is shown in figure 2.26, where a n-type symmetrical one has been considered as input stage.

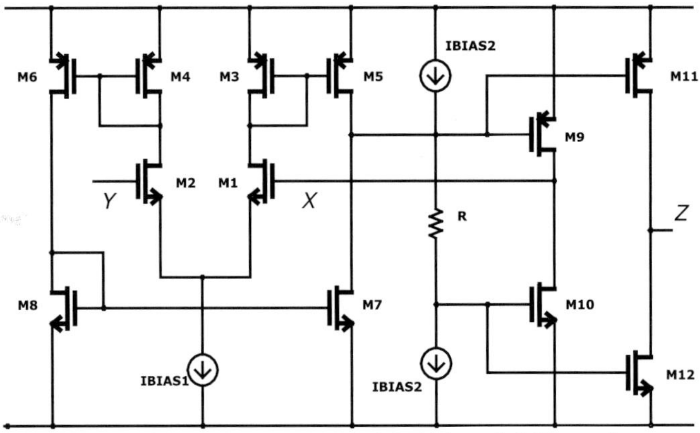

Figure 2.26– N-type symmetrical OTA-based CCII topology

The value of α parameter is very close to eq. (2.55), evaluated for the previous OTA-based CCII, in fact:

$$\alpha = \frac{V_X}{V_Y} \cong \frac{\frac{r_{05}\,r_{07}}{r_{05}+r_{07}}\left(\frac{g_{m5}}{2g_{m3}}+\frac{g_{m6}g_{m7}}{2g_{m4}g_{m8}}\right)\frac{r_{09}\,r_{010}}{r_{09}+r_{010}}(g_{m9}+g_{m10})\,g_{m2}}{1+\frac{r_{05}\,r_{07}}{r_{05}+r_{07}}\left(\frac{g_{m5}}{2g_{m3}}+\frac{g_{m6}g_{m7}}{2g_{m4}g_{m8}}\right)\frac{r_{09}\,r_{010}}{r_{09}+r_{010}}(g_{m9}+g_{m10})\,g_{m1}} \cong \frac{g_{m2}}{g_{m1}} \tag{2.60}$$

The current parameter β is given by eq.(2.61).

$$\beta = \frac{I_Z}{I_X} = \frac{(g_{m11}+g_{m12})}{(g_{m9}+g_{m10})}\,\frac{\frac{r_{011}r_{012}}{r_{011}+r_{012}}}{\frac{r_{011}r_{012}}{r_{011}+r_{012}}+R_{ZLOAD}}\,\frac{\frac{r_{09}r_{010}}{r_{09}+r_{010}}+R_{XLOAD}}{\frac{r_{09}r_{010}}{r_{09}+r_{010}}} \cong \frac{g_{m11}+g_{m12}}{g_{m9}+g_{m10}} \tag{2.61}$$

Eq. (2.16) still gives the value of Z_Y parasitic impedance, while Z_X parasitic impedance resistive component is given by:

$$R_X = \frac{1}{\frac{r_{09}+r_{010}}{r_{09}\,r_{010}}+\left(\frac{g_{m6}}{2g_{m4}}+\frac{g_{m5}g_{m8}}{2g_{m3}g_{m7}}\right)\frac{r_{05}\,r_{07}}{r_{05}+r_{07}}(g_{m9}+g_{m10})g_{m1}} \cong \frac{1}{B\,\frac{r_{05}\,r_{07}}{r_{05}+r_{07}}(g_{m9}+g_{m10})g_{m1}} \tag{2.62}$$

being B the mirror ratio of the symmetrical OTA. An inductive behaviour will result again, and this inductive component is:

$$L_X = \frac{1/p_0}{\frac{r_{09}+r_{010}}{r_{09}r_{010}} + \left(\frac{g_{m5}}{2g_{m3}} + \frac{g_{m6}g_{m7}}{2g_{m4}g_{m8}}\right)\frac{r_{05}r_{07}}{r_{05}+r_{07}}(g_{m9}+g_{m10})g_{m1}} \cong \frac{1/p_0}{B\frac{r_{05}r_{07}}{r_{05}+r_{07}}(g_{m9}+g_{m10})g_{m1}} \tag{2.63}$$

The choice of a suitable B can improve the parasitic impedance at X node. The output impedance at Z node is high, thanks to the chosen topology, and given by the parallel of r_{011} and r_{012}.

$$Z_Z = \frac{r_{011}r_{012}}{r_{011}+r_{012}} \tag{2.64}$$

Table 2.11 summarises the CCII parameters for the circuit shown in figure 2.26.

Current Conveyor Characteristics				
Data	**Value**	**Parameter**	**Theoretical Value**	**Simulated Value**
Voltage Supply	±0.75 V	**Voltage Gain (α)**	1.0007	0.9999
Power Consumption	120 μW	**Current Gain (β)**	1.00	1.00 (Rx=Rz=10KΩ)
3dB Bandwidth	17 MHz	**Node Y Parasitic Impedance**	0.2 pF	0.195 pF
Dynamic Range	-350mV, +750mV	**Node X Parasitic Resistance**	23.8 Ω	24.7 Ω
Biasing Current IBIAS1	10 μA	**Node X Parasitic Inductance**	30 μH	60 μH
Current in M9,M10	10 μA	**Node Z Parasitic Impedance**	1.43 MΩ	1.29 MΩ

Table 2.11 –CCII characteristics of the circuit shown in figure 2.26

Obviously a p-type symmetrical OTA based CCII can be also implemented, as reported in figure 2.27.

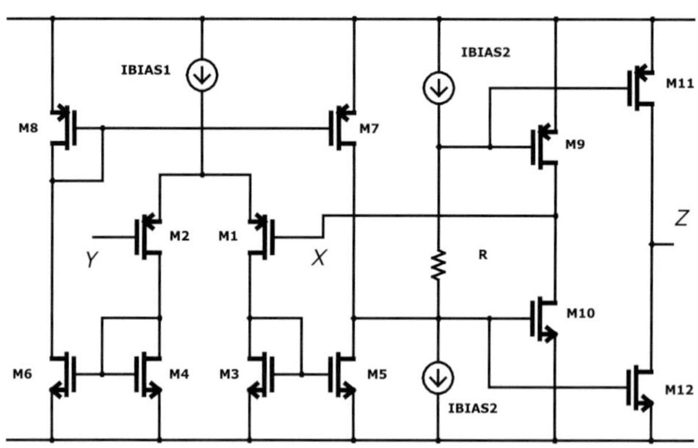

Figure 2.27– P-type symmetrical OTA-based CCII topology

The theoretical expressions are exactly the same seen for the n-type one. We present only the overall results in table 2.12.

Current Conveyor Characteristics				
Data	**Value**	**Parameter**	**Theoretical Value**	**Simulated Value**
Voltage Supply	±0.75V	**Voltage Gain (α)**	1.017	0.9994
Power Consumption	120 μW	**Current Gain (β)**	1.00	1.00 (Rx=Rz=10KΩ)
3dB Bandwidth	15 MHz	**Node Y Parasitic Impedance**	0.5 pF	0.47 pF
Dynamic Range	–750mV, +300mV	**Node X Parasitic Resistance**	25.8 Ω	27.5 Ω
Biasing Current IBIAS1	10 μA	**Node X Parasitic Inductance**	33.5 μH	68 μH
Current in M9,M10	10 μA	**Node Z Parasitic Impedance**	1.43 MΩ	1.29 MΩ

Table 2.12 – CCII characteristics for the circuit shown in figure 2.27

The two previous topologies can be joined to obtain a rail-to-rail CCII, as shown in figure 2.28. When the input voltage applied to Y node is high, the n-type based CCII operates, while the p-type one works for low inputs. In this situation all the theoretical expressions presented before are still valid. Considering the rail-to-rail solution, the theoretical analysis becomes quite complicated and has not been reported here. If a good matching between the OTAs is ensured, the expressions from (2.60) to (2.64) can be used to evaluate the CCII behaviour.

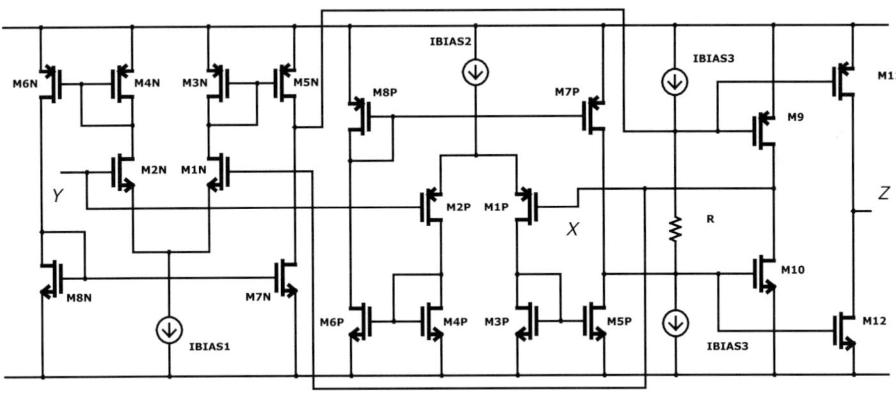

Figure 2.28– Rail-to-rail symmetrical OTA-based CCII topology

The comparison between theoretical and simulated results is presented in table 2.13. It has to be noted that a true rail-to-rail behaviour has been obtained. This important result has been confirmed by experimental measures presented in the Appendix B.

Current Conveyor Characteristics				
Data	Value	Parameter	Theoretical Value	Simulated Value
Voltage Supply	±0.75 V	Voltage Gain (α)	0.994	1.0007
Power Consumption	180 μW	Current Gain (β)	1.00	1.00 (Rx=Rz=10KΩ)
3dB Bandwidth	14 MHz	Node Y Parasitic Impedance	0.7 pF	0.67 pF
Dynamic Range	-750mV - +750mV	Node X Parasitic Resistance	20.2 Ω	20.8 Ω
Current IBIAS1, IBIAS2	10 μA	Node X Parasitic Inductance	49 μH	63 μH
Current in M9,M10	10 μA	Node Z Parasitic Impedance	1.43 MΩ	1.27 MΩ

Table 2.13 – CCII characteristics for the circuit shown in figure 2.28

In the previous pages several different solutions for the implementation of a CCII have been introduced. Theoretical and simulated results have been summarised in the above tables. If a CCII is needed for a certain application its topology can be chosen accordingly to these results. Starting from the first topology, reported in figure 2.14, we can say that such a solution allows to obtain wider bandwidth, but it is not suitable for LV application, being the typical supply voltage equal to ± 1.5 V. Moreover it is not suitable for LP application too, because of its impedance levels and a very low X node impedance can be obtained only with biasing currents that affect in a dramatic way the overall power consumption.
To overcome these limitations, differential pair based CCII have been proposed, the first of which is reported in figure 2.15. As shown before, their implementation is nothing more than the application of the theory presented in the first chapter. The solution shown in figure 2.15 is really suitable for LV LP applications, but the resulting impedance levels are not acceptable (for example, node X parasitic impedance is about 15 KΩ). The feedback topology has been improved, as in figure 2.16, in order to overcome this problem. The impedance levels obtained are now really good (now Rx is about 25 Ω), the power consumption is limited (58 μW), but the overall topology does not seem to be a true LV one (±1 V supply). That's why the "current sensing" technique has been modified, leading to the circuit shown in figure 2.18. The only limitation is now represented by the fact that a class A solution have been implemented. In a very rough way, this circuit has been modified in the one presented in figure 2.19, which shows a class AB behaviour. The results reported in table 2.5 are very good, but they are based on a given biasing current for the output stage. Due to technological spread, the biasing of the class AB stage reported in figure 2.19 may vary in an unacceptable way. Different output stages are required, in order to design a more robust circuit. The first idea is to join together the two different solutions represented by the circuit in figure 2.14 and 2.19. The resulting circuit

has been reported in figure 2.20. All the CCII characteristics are satisfactory, but once more a circuit unsuitable for LV application has been obtained (typical supply voltage ±1.5 V). Replacing the output stage with a LV one [18], circuits presented in figure 2.21 and 2.22 have been derived.
The only drawback that they exhibit is an unbalanced dynamic range, that is the nMOS based topology has a dynamic range towards the positive rail, while that of pMOS based one is towards the negative rail. These two characteristics may be joined, designing the circuit presented in figure 2.23, representing a rail-to-rail current conveyor.
The theory presented in the first chapter has been once more confirmed designing the circuit of figure 2.25, an OTA based CCII, for which some performance are better than those obtained for the circuits in figure 2.21 and 2.22, especially for what concerns the dynamic range.
Analog considerations, about the circuits shown in figures 2.21-2.23, can be done considering, as CCII input stage, a symmetrical OTA. Also in this case, we have presented n-type, p-type and rail-to-rail topologies (figures 2.26, 2.27 and 2.28). The last one has very good performance and can be considered, without doubts, the best solution of general CCII illustrated in this chapter.

2.3 THE CCII: STATE OF THE ART AND FUTURE TREND

The first topologies of CCIs and CCIIs were proposed more than thirty years ago, but their high performance and high precision applications continue to be of a particular interest. This is why in many situations the CCII-based solution can reduce the system complexity with respect to more traditional implementations that utilise commercial integrated circuits as the voltage OAs.
Since that time, hundreds of presentations and articles have witnessed the evolution of the first CC concept, demonstrating the universality of the element in the synthesis of almost all the active functions.
Maintaining the matrix characteristics, CCII evolution has regarded its internal topology, so to enhance the performance and then the utility of the CCII block. Certainly, the first implementations of integrated CCIIs presented in the literature or as commercial products were realised in bipolar technology. The advent of MOS transistors, however, has pushed the most of actual integrated solutions (in particular, those related to mixed analog - digital electronics)

towards the use of CMOS technology, which shows a higher design simplicity and also a very low power consumption, particularly attractive in low-voltage portable-system applications. Even if CMOS devices suffer from other problems, such as body effect, threshold voltage mismatch between nMOS e pMOS and lower Gm values, its low cost especially for a standard technology represents a decisive feature towards its success and utilisation.

The use of CC in the implementation of analog basic function is very spread. With respect to commercial solutions with OAs, CCII-based ones are often simpler in structure and more versatile, especially for those circuits which involve high impedance current output capabilities. Furthermore, through the use of CCIIs, the heavy OA limitation of the constant gain-bandwidth product is definitely overcome.

With basic CCIIs, it is also possible to implement a number of analog applications such as : voltage and current amplifiers, current differentiators and integrators, capacitance multipliers, impedance simulators and converters, biquadratic filters, voltage-to-current and current-to-voltage converters, instrumentation amplifiers, oscillators and waveform generators, etc..

Many solutions have been presented in the past but new possibilities can be certainly considered for the future, especially at *lower supply voltages* and with *more reduced dissipation*, as will be explained in the next chapter.

An important aspect to study can be the implementation of an *"ideal"* conveyor, with the improvement of its non-ideal characteristics (input and output real and imaginary parasitic impedances, current and voltage transfer functions, etc.) also through the development of novel architectures or the utilization of circuit tricks. This goes in the direction of an *enhanced model* for the CCII device [19].

Moreover, the development of novel architectures able to implement analog functions with a *lower number of basic elements* (for example, the design of a biquadratic filter through the use of a single CCII) can be of certain interest for the researchers. Also new fields of applications for the CCIIs, as for example, the development of *current-mode sensor interface* circuits (see chapter 5), can be explored.

Finally, it certainly deserves a deeper look also the study about the generalisation of the CCII structure towards *differential and fully differential solutions* with the basic block formed by more than three terminals (see chapter 4). These circuits, in fact, seem particularly appreciated in some applications as, for example, four-quadrant multipliers and filters and also allow to revisit the proposed functions in a differential way.

References.

[1] A. S. Sedra, G. W. Roberts. "Current conveyor theory and practice". In *Analogue IC design: The current mode approach.* Peter Peregrinus, London, 1990.

[2] A. Sedra, K. C. Smith. A second generation current conveyor and its applications. IEEE Transactions on Circuit Theory. CT-17; 1970; pp. 132-134.

[3] I. A. Awad, A. M. Soliman. New CMOS realization of the CCII-. IEEE Transactions on Circuit and Systems-I. nr. 4; vol. 46; 1999; pp. 460-463.

[4] M. C. H. Cheng, C. Toumazou. 3V MOS current conveyor cell for VLSI technology. Electronics Letters. nr. 3; vol. 29; 1993; pp. 317-318.

[5] H. W. Cha, K. Watanabe. Wideband CMOS current conveyor. Electronics Letters. nr. 14; vol. 32; 1996; pp. 1245-1246.

[6] A. Fabre, M. Alami. A precise macromodel for second generation current conveyors. IEEE Transactions on Circuit and Systems-I. nr. 7; vol. 44; 1997; pp. 639-642.

[7] X.Zhang, X.Ni, M.Iwahashi, N.Kambayashi. Realization of universal active complex filter using CCIIs and CFCCIIs. Analog Integrated Circuits and Signal Processing. nr. 20; 1999; pp. 129-137.

[8] A. Fabre, O. Saaid, H. Barthelemy. On the frequency limitations of the circuits based on second generations current conveyors. Analog Integrated Circuits and Signal Processing. nr. 7; 1996; pp. 113-129.

[9] C.A. Karyababkas, C.A.Papazoglou. Low sensitive CCII-based biquadratic filters offering electronic frequency shifting. IEEE Transactions on Circuits and Systems II. nr.5; vol.46; 1999; pp.527-538.

[10] H.Barthelemy, G.Ferri, N.Guerrini, A 1.5 V CCII-based tunable oscillator for portable industrial applications, Proceedings of International Conference on Industrial Electronics, 2002. L'Aquila, Italy.

[11] G. Palumbo, S. Palmisano, S. Pennisi. *CMOS current amplifiers*. Kluwer Academic Publishers, Boston, 1999.

[12] G.Ferri, P.De Laurentiis, G.Stochino. Current conveyors II. Electronics World; April 2001; pp.300-302.

[13] G.Ferri, N.Guerrini. High valued passive element simulation using low-voltage low-power current conveyors for fully integrated applications. IEEE Transactions on Circuits and Systems II. n.4; vol.48; 2001; pp.405-409.

[14] G.Ferri, N.Guerrini. Low-voltage low-power current conveyors : design and applications. Alta Frequenza. nr.4; vol. 12; 2000; pp.59-63.

[15] G.Ferri, N.Guerrini. Low-voltage low-power novel CCII topologies and applications. Proceedings of the IEEE International Conference on Electronic Circuits and Systems, 2001; Malta.

[16] G.Ferri, N.Guerrini, M.C.Piccirilli. Low voltage current conveyor-based universal biquad filter. Proceedings of International Conference on Industrial Electronics, 2002. L'Aquila, Italy.

[17] T. Kurashina, S. Ogawa, K. Watanabe. A high performance class AB current conveyor. Proceedings of the IEEE International Conference on Electronic Circuits and Systems, 1998. Lisboa, Portugal.

[18] A.Torralba, private discussion, L'Aquila, July 2002.

[19] A. Arbel. Negative Feedback Revisited. Analog Integrated Circuits and Signal Processing. nr. 10; 1996; pp. 157-178.

CHAPTER III

CCII LOW VOLTAGE LOW POWER DESIGN AND CHARACTERISTICS

In this chapter, some circuit solutions and a design example for the implementation of low-voltage low-power (LV LP) CCIIs in CMOS technology will be investigated. The behaviour of the topologies presented in the previous chapter will be analysed at reduced supply voltages and new LV LP topologies will be also presented. An analysis of two important parameters in LV LP applications, offset and noise, will conclude the chapter.

3.1 DESIGN TECHNIQUES FOR LOW VOLTAGE LOW POWER (LV LP).

Recently, integrated circuit designers have been putting an increasing effort into the reduction of supply voltage and power dissipation of digital, analog and mixed circuits and systems. This is mainly due to the following factors. First of all, the need to reduce power consumption in modern high-density digital systems. As chip components get closer together, the problem of heat dissipation increases while break-down voltages of the components on chip reduce as geometries become smaller. Secondly, the explosive growth of the market of portable battery-operated electronics, which has stimulated the demand for LP topologies able to operate at reduced supplies and to ensure a longer battery lifetime. In this sense, LV LP techniques are mandatory for implantable devices, such as pace-makers, blood flow-meters, hearing instruments, and auditory stimulators, and for portable radios and PCs, cellular phones, pagers, etc.. Thirdly, LP solutions are spreading into fields like filters, audio signal processing, EMC compliant systems, etc, where non-invasive sensors can be integrated on-chip. Lastly, there is a huge scientific interest in exploring the technological and physical limits of integrated devices, taking into account the continuous technology scaling of circuit dimensions, which increases both speed and density of digital architectures and the electrical field related to a constant supply voltage [1,2,3,4].

Digital circuits have always characterised the reference supply voltage value, which has scaled from 5 V to 3.3 V, and lower. In digital cells, power reduction can be obtained easily through the lowering of the supply voltage. In fact, current consumption in CMOS circuits is directly proportional to the square of the supply voltage.

Analog signal processing has always had a huge importance in real-time signal processing because of the analog kind of the physical phenomena and of all the signals which come from and go to the outside world. This justifies a strong and continuous interest in analog system design.

LV analog design techniques differ considerably from traditional supply design and the basic analog blocks have to be reconsidered in a LV environment. LV requirements can be summarised as follows. Especially for portable applications, LV circuits need to be compatible with common battery voltage values. Then, they have to implement the required analog functions with sufficiently good performance. In this sense, traditional architectures available for working at low supply rails are generally inadequate as well as typical models for transistors which have to be implemented with a new particular attention in the boundary region between weak and strong inversion, where transistors are often biased.

In all the basic blocks, as the operational amplifier, the new constraints concern both the full input swing (performed through the use of two complementary pairs in parallel) and the full output one (so to complete the rail-to-rail operation; this last performance is obtained through a class-AB stage with low output quiescent current and output current control). Amplifier input stages have also to show a transconductance independent from the input common mode voltage, so to present the same circuit characteristics in any biasing condition. As a sum of all these factors, we can say that *in LV design it is fundamental an efficient use of the supply voltage range*.

In the literature, a circuit can be included in the LV category according to the number of stacked gate-source (threshold) and drain-source (saturation) voltages. There is not a predefined value which exactly determines the boundary between a non-LV and a LV topology. The term LV can be typically used for circuits that are able to operate at a supply voltage of $2\ V_{TH} + 2\ V_{DSAT}$, while very low voltage (VLV) circuits have also to work at only $V_{TH} + V_{DSAT}$. Of course, this is only a possible definition but, in this sense, numerical supply values are strictly related to the technology used and tend to decrease during the years with the scaling of circuit sizes.

In analog circuits, the reduction of the supply voltage does not necessarily corresponds to a decrease of related power consumption. In this case, the "folding" technique can replace the traditional "stacking" of transistors. To keep the power low, analog circuits have to be designed as much simple as possible.

Moreover, a trivial decrease of biasing currents, which can reduce circuit dissipation, degrades circuit performance, first of all bandwidth and dynamic range (considering, for example, operational amplifiers, where also minimum size transistors cannot be used because of gain, noise and offset voltage constraints). As a consequence, chip area cannot be drastically reduced with the lowered feature dimensions.
Power limitations are mainly linked to: parasitic capacitances; traditional current-inefficient amplifiers, not optimised for a low quiescent dissipation; peak-to-peak limitations. As a result, *LP design is characterised by an efficient use of the supply current*, through the utilisation of class-AB output stages and an efficient frequency compensation strategy.
The combination of these constraints and requirements gives the basic rules to be followed to design LV LP circuits, even if in a large number of analog applications, and also for what our study has concerned, designers focus their studies on the development of topologies able to work at reduced supply. In this case, using typical values of biasing current in the μA range, circuits show generally a reduced power consumption (for example, not higher than 1 mW) [5].
Concerning the technology, the continuous reduction of the threshold voltage in CMOS has definitively directed LV design towards CMOS itself, which is also typically characterised by a very low quiescent power consumption. Reducing the supply voltage, CMOS transistor is often biased to work in weak inversion region: in this sense, the use of good transistor models is of a fundamental importance [6].
Finally, we want to remark that, for LV LP applications, the current-mode approach can be a valid alternative to traditional circuits to obtain high performance architectures, because the designer is concerned with current levels for circuit operation instead of node voltages. In this manner, as well known, the gain-bandwidth product limitation, typical of operational amplifiers, is overcome.

3.1.1 LV LP biasing current design.

All the CCII-based topologies introduced in the previous pages have a minimum operating supply voltage related to the drain-source (saturation) voltage required by the biasing transistors, which has to be minimised to reduce the circuit supply voltage. For this reason, in this paragraph, the implementation of suitable biasing current generators is presented.

Figure 3.1a shows the basic topology of a traditional nMOS current mirror, performing also an inverting current amplifier. A complementary circuit can be easily developed for the pMOS type, as pictured in figure 3.1b.

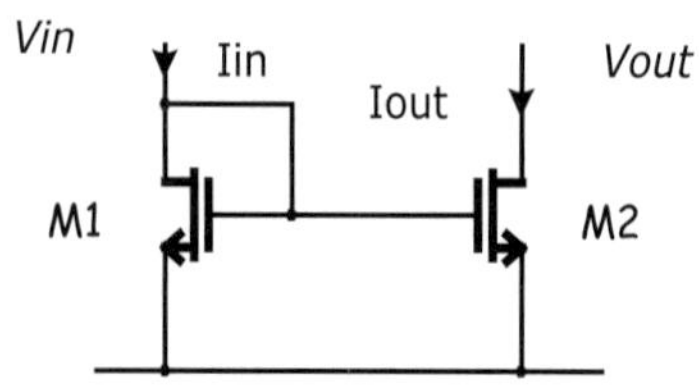

Figure 3.1a – nMOS current mirror

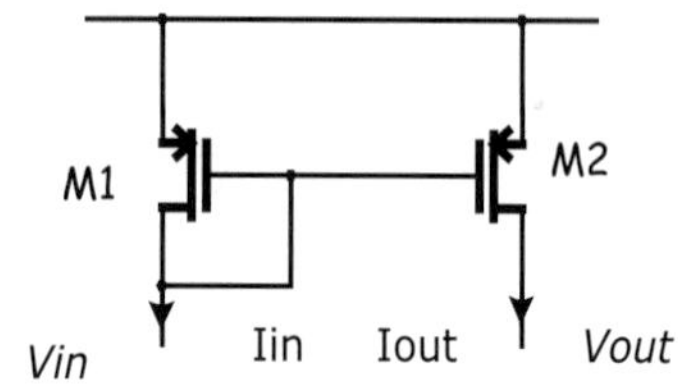

Figure 3.1b – pMOS current mirror

The basic current mirror topology is the same if transistors work either in saturated or subthreshold region (this condition may easily be verified in LV circuits), even if in the last case the frequency response is very bad and linearity is quite poor for low V_{DS}. Typically, the output current I_{OUT} is approximately related to I_{IN} according to the transistor W/L ratios. The output impedance, determined at M2 drain, is not particularly high, and also is affected by channel length modulation, being equal to MOS transistor output resistance r_0. For this current mirror we can write the following relationships (see figures 3.1a and 3.1b) :

$$\begin{cases} V_{IN}\min \cong V_{TH} \\ V_{OUT}\min = V_{DSAT} \\ R_{OUT} = r_{02} \end{cases} \tag{3.1}$$

In fact, due to MOS output characteristics, only when the output voltage is greater than the saturation voltage, this transistor behaves like a current generator. There is a minimum limit also for V_{IN}, given by the threshold voltage V_{TH} : in this case M1 works in the subthreshold region. Moreover, the finite output resistance, due to channel length modulation effect, gives some differences between I_{IN} and I_{OUT} if V_{IN} is different from V_{OUT}.

To obtain higher impedances, it is possible to utilise a cascode structure, as shown in figure 3.2, where two MOSFETs have been placed one above the other [7].

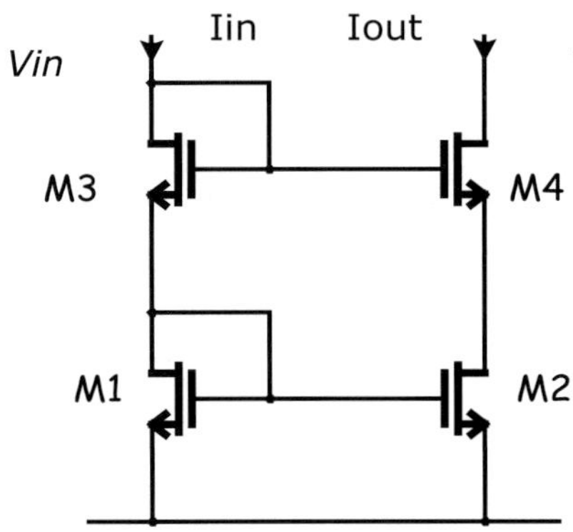

Figure 3.2 – nMOS cascoded current mirror

The use of this topology can increase the circuit voltage gain, but also reduces the output signal swing, so preventing its utilisation in LV architectures. For the current mirror shown in Figure 3.2, we can write the following relationships:

$$\begin{cases} V_{IN}\min = 2V_{TH} \\ V_{OUT}\min = 2V_{DSAT} \\ R_{OUT} \approx g_{m4} r_{04} r_{02} \end{cases} \qquad (3.2)$$

The solution reported in Figure 3.2 allows to improve only the output impedance level, while shows worst characteristics with respect to the basic mirror topology (figure 3.1) in terms of minimum operating voltage, given by 2 V_{TH}.
An enhanced current mirror capable of operating with lower supply voltages or performing higher swings is reported in figure 3.3 [7].

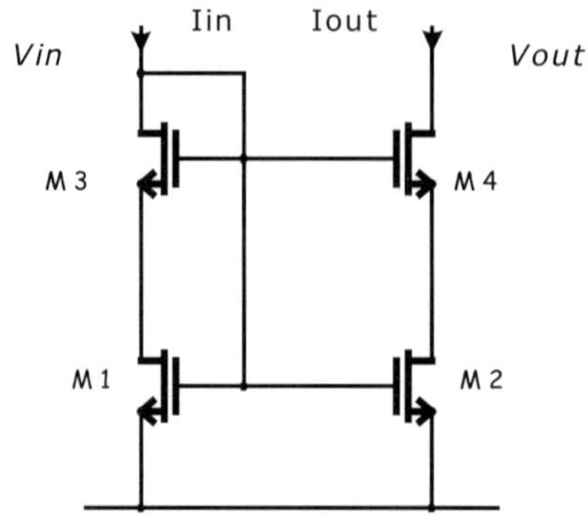

Figure 3.3 – Low voltage nMOS cascoded current mirror

In this case, the minimum operating supply voltage has been decreased, while the output resistance is still very high.

$$\begin{cases} V_{IN}\ \min = V_{TH} \\ V_{OUT}\ \min = 2\,V_{DSAT} \\ R_{OUT} \approx g_{m4} r_{04} r_{02} \end{cases} \tag{3.3}$$

Unfortunately, all these solutions proposed for current generators heavily depend on the supply voltage value. In fact, from figure 3.4, we have that the input current, which is mirrored into the biasing current, given by:

$$I_{IN} = \frac{V_{DD} - V_{GS1}}{R} \tag{3.4}$$

is strongly dependent on the supply voltage value V_{DD}. This means that if this value reduces (for example, if the supply voltage is performed by a single-cell battery that can discharge), all the biasing currents reduce too and circuit performance consequently degrade.

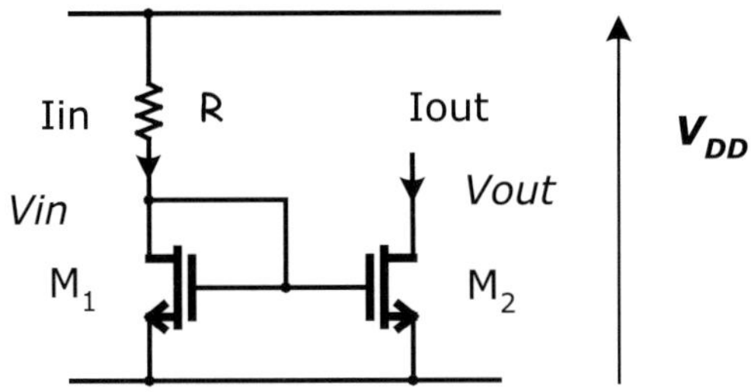

Figure 3.4 – Complete nMOS current mirror

In all the portable system applications, where the supply voltage is a single-cell battery, it is possible to use a different topology for biasing current, which is independent from the supply voltage and is only related to the MOS threshold voltage (which is a typical value for each technology). This circuit is shown in figure 3.5 (its complementary version is reported in [8]).

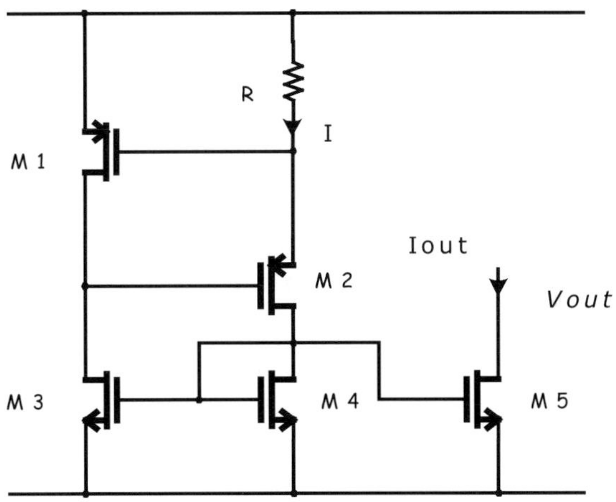

Figure 3.5 – Low voltage current source (independent from supply value)

From the analysis of the circuit depicted in figure 3.5 it is possible to write:

$$V_{GS1} = RI = V_{TH,P} + \sqrt{\frac{2I}{K_P\left(\frac{W}{L}\right)_1}} \tag{3.5}$$

If M1 aspect ratio is very high, the previous expression can be simplified as follows:

$$V_{GS1} = RI \cong V_{TH,P} \tag{3.6}$$

from which:

$$I \cong \frac{V_{TH,P}}{R} \tag{3.7}$$

and consequently:

$$I_{OUT} \cong \frac{V_{TH,P}}{R} \frac{\left(W/L\right)_5}{\left(W/L\right)_4} = \frac{V_{TH,P}}{R} \tag{3.8}$$

for unitary gain current mirrors. Using this circuit, suitable for LV applications (the minimum operating supply voltage is $2V_{TH} + V_{DSAT}$), the biasing current is independent from the supply voltage and, consequently, insensitive to any supply discharge.

3.2 LOW VOLTAGE LOW POWER CCII DESIGN

Several CCIIs topologies presented in the previous chapter are based on a differential pair followed by an output stage. Theoretical analysis and simulations have confirmed that this solution ensures good performance also in terms of LV LP characteristics. In order to give a sort of guideline in the design of such a current conveyor, the design flow of CCII topology will be presented. It is important to point out the fact that in this section we don't want to show how particular performance can be obtained, but only give an example of a LV LP CCII design. Let us consider the class AB CCII, based on a differential pair (figure 2.19), here represented again in figure 3.6.

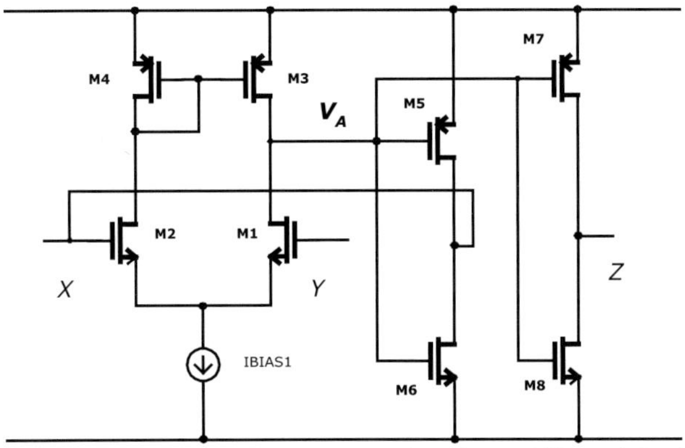

Figure 3.6 – Class AB CCII topology based on a differential pair

The first design step concerns the differential pair, pictured in figure 3.7.

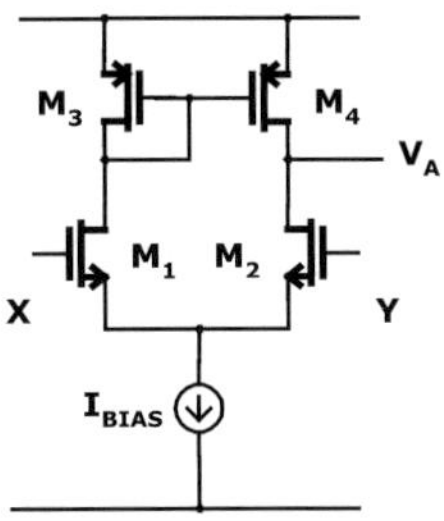

Figure 3.7 – Differential pair

For LV LP applications, a special care has to be used in the design of current sources. In this sense, we have to remember that at these supply values it is fundamental to guarantee signal dynamic and a constant bias value, independent from supply discharge (see figure 3.5).
The choice of I_{BIAS} current is very important, because its value affects, in particular:

- Power consumption
- Differential pair transistors sizes
- X node parasitic impedance

Once the current source topology has been implemented, M1-M4 sizes have to be set. From eq.(3.9), it is clear that the error due to input voltage buffer is strictly related to g_{m1} and g_{m2} values that have to be equal, in fact:

$$a = \frac{V_X}{V_Y} \cong \frac{g_{m2}}{g_{m1}} \tag{3.9}$$

For what concerns the current transfer β, the following formula has been derived, if the load resistances are not very high when compared to the MOS output resistance value.

$$\beta = \frac{I_Z}{I_X} \cong \frac{g_{m8} + g_{m7}}{g_{m6} + g_{m5}} \tag{3.10}$$

Let us assume a 10 μA biasing current for the differential pair. As a consequence, M1 and M2 have a stand-by current equal to 5 μA. If the ratio W/L is set equal to 150, the g_{m1} and g_{m2} values are, in the chosen technology (AMS 0.5 μ), both 138 μA/V.
M3 and M4 sizes have to be set considering also the output stage design. In fact, their value directly affects the X node offset, because they set V_A that is the gate biasing voltage of the output stage.
M5 and M6 sizes are also important because the corresponding transconductance value affects the X node parasitic resistance, as follows:

$$R_X = \frac{1}{\frac{r_{05} + r_{06}}{r_{05} r_{06}} + g_{m2} \frac{r_0}{2}\left(g_{m5} + g_{m6}\right)} \cong \frac{2}{g_{m2} r_0 \left(g_{m5} + g_{m6}\right)} \tag{3.11}$$

Let us impose also an output biasing current equal to 10 μA. Using the g_m equations, it could be possible to determine output transistors and active loads sizes. The output transistor sizes have been set to 13μm/2μm for the n-type MOS and 13μm/1μm for the p-type ones. In order to minimise the output voltage offset the active load size ratios has been chosen equal to 7μm/1μm. The simulated residual offset is equal to 37.5 μV.
At a first glance this does not appear a rigorous approach to CCII design, but with this simple example we would like to concentrate the reader attention on the following particular aspects:

- it is not very easy to separate the different design choices;
- each CCII characteristic is affected by several design constraints;
- design settings come out after a trade-off among opposite needs;
- *but* the first and most important design step is represented by choice of the topology.

As an example, from the previous discussion it is clear that differential pair biasing and aspect ratio, active load sizes and output stage biasing are all involved in the offset and X node parasitic impedance evaluation, dynamic range and so on. Modifying only one parameter to obtain a specific goal, as power consumption, may compromise other performance, as offset, for example. A good approach may be represented by a preliminary synthesis that leads to a rough circuit design, followed by a carefully analysis of the designed solution, to check the affordability of the results found.
Let us consider the following figures, where we present the results of different simulations, obtained through PSPICE, on the current conveyor proposed in figure 3.6, and designed as previously indicated.
In these simulations, the biasing current has been implemented through the current mirror of figure 3.5.
Firstly, the input voltage buffer behaviour has been investigated. A DC sweep simulation has been performed, to check the range where the voltage on X node is equal to the one applied to Y node. The resulting graph is something like the one reported in figure 3.8.

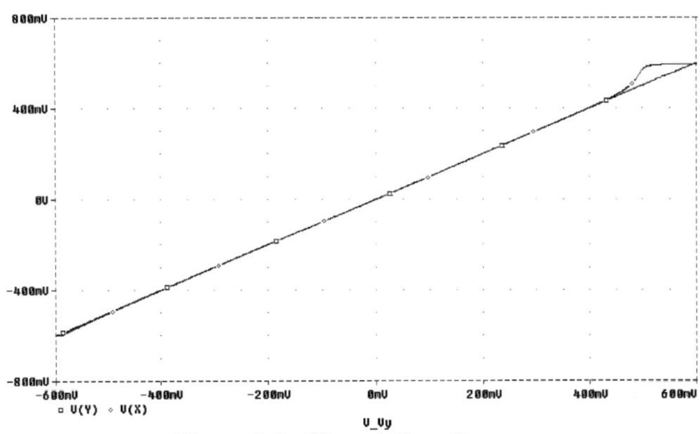

Figure 3.8 – V_Y and V_X voltages

The range for which $V_X=V_Y$ is limited, except for rail-to-rail topologies, where it is almost equal to the difference between positive and negative supplies. For the circuit simulated shown in figure 3.6, this range is approximately 800 mV, which represents the 67 % of the total supply. The offset between X and Y node voltages is shown in figure 3.9, where the difference between V_X and V_Y has been pictured.

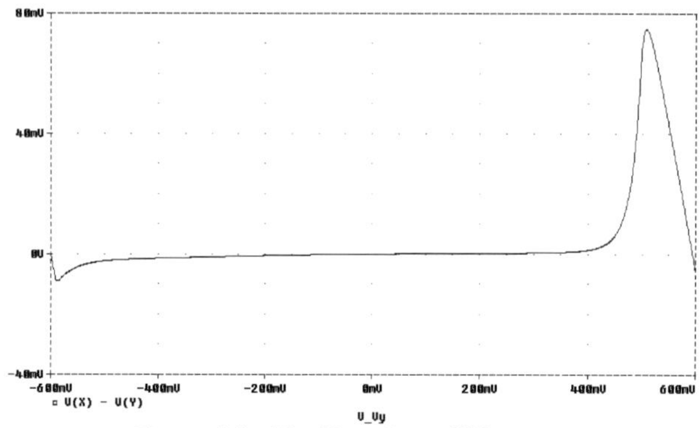

Figure 3.9 – V_X–V_Y voltage difference

The operating voltage range can be chosen fixing a maximum error on the voltage transfer between X and Y nodes, so it can be useful a more detailed view of the graph reported in figure 3.9, shown in figure 3.10.

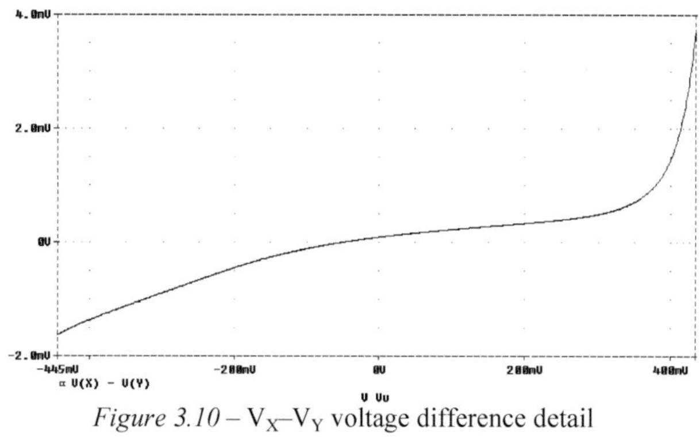

Figure 3.10 – V_X–V_Y voltage difference detail

Moreover, it is interesting to check the results found with the previous simulations performing an analysis in the time domain, as in figure 3.11. In this picture, it is possible to compare two signals, an input sinusoidal voltage having an amplitude of 400 mV and a frequency of 1 KHz, applied to Y node, and the resulting signal on X node. It is clear that they are very similar, thus confirming the fact that the operating range is not lower than 800 mV.

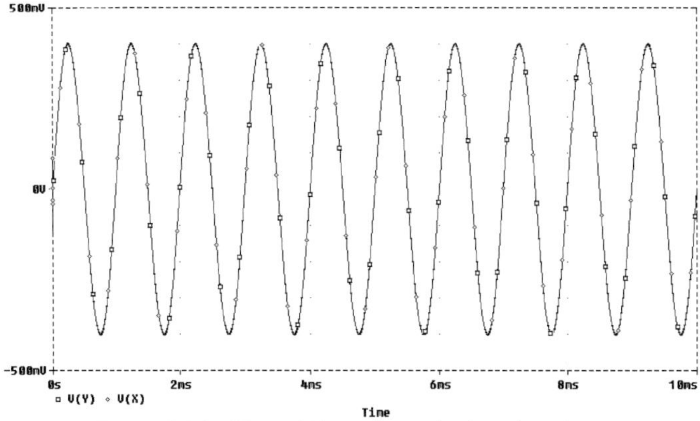

Figure 3.11 – V_X and V_Y voltages in time domain.

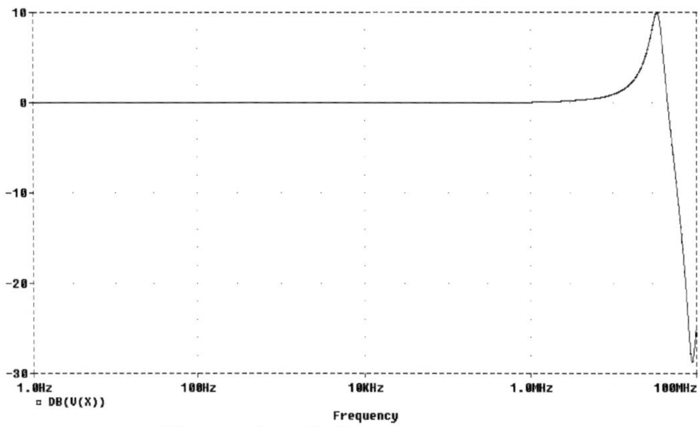

Figure 3.12 – V_X frequency response

Then, we would like to investigate what is the frequency range for the designed CCII, so an AC simulation has been performed. The resulting graph has been pictured in figure 3.12.
The frequency behaviour reported in figure 3.12 suggests that, for high frequency applications, a compensation is needed, so a Miller capacitor has to be connected between gate and drain of M5 transistor.
All the previous simulations have been performed without connecting any load to X node. In order to estimate the performance of the designed CCII, the effects of an external load at X node have also to be considered.
In figure 3.13 the voltages at X and Y nodes, when a 10 KΩ resistance is connected to X node, are presented. It can be seen that the range is not reduced with respect to the unloaded case, so the considered load does not affect the CCII behaviour. Obviously, using different load values, this will not be true. In fact, connecting to X node a 2 KΩ resistance, the dynamic range heavily decreases, as shown in figure 3.14.
Let us now check the current buffer characteristic of the designed CCII. This control is performed simply connecting two loads to the two current nodes, that are X and Z. Once that a voltage has been applied to Y node, it is possible to verify the current flowing in the two external loads. The result is shown in figure 3.15, where two equal resistances valued 2 KΩ have been used as X and Z loads.

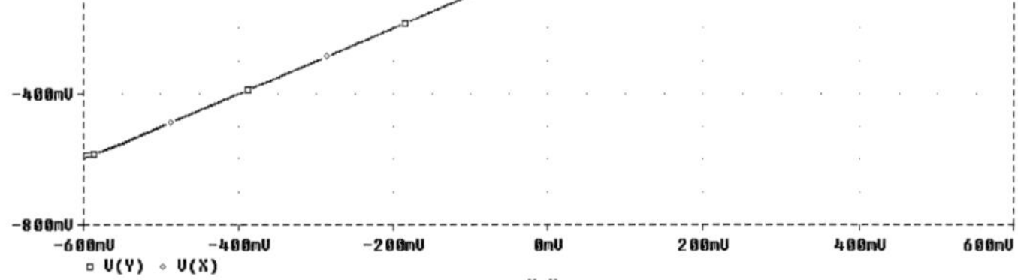

Figure 3.13 – V_Y and V_X voltages with a 10 KΩ resistance connected to X node

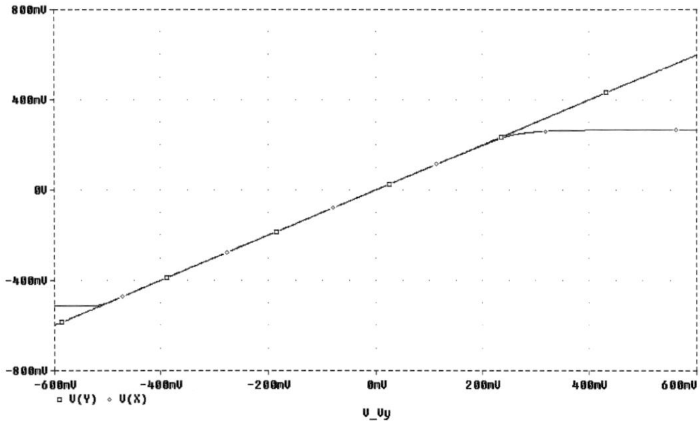

Figure 3.14 – V_Y and V_X voltages with a 2 KΩ resistance connected to X node

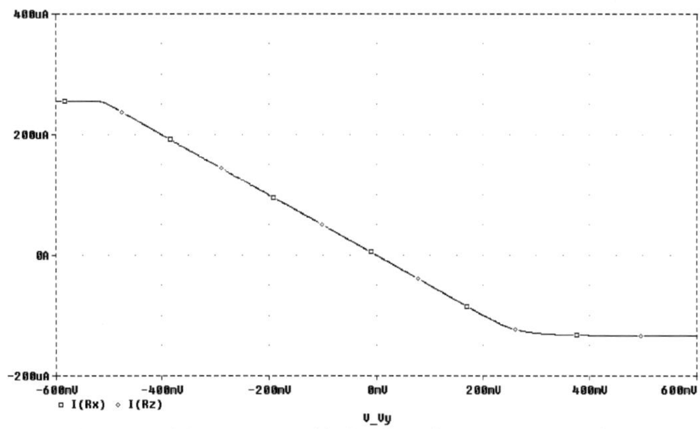

Figure 3.15 – I_Z and I_X currents with 2 KΩ resistance connected to X node

As for the voltage transfer characteristic, simulations in time domain may help to understand if the final circuit will work properly or not. The result for 1 KHz sinusoidal wave is shown in figure 3.16. The two currents I_Z and I_X are very close one each other.

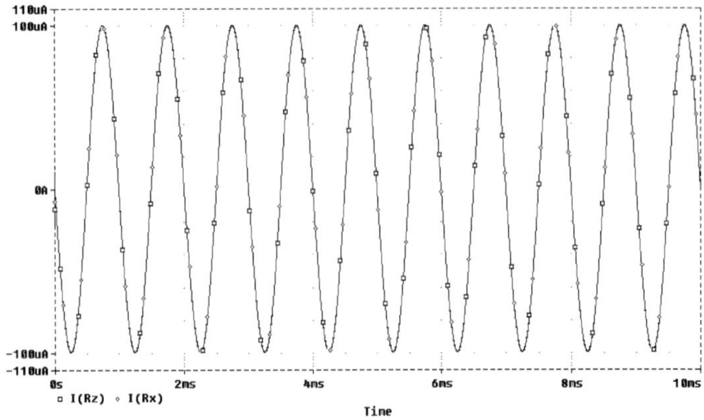

Figure 3.16 – I_Z and I_X currents in time domain

Finally, CCII parasitic impedances have to be estimated. This can be done applying a suitable signal source at the node to be investigated and performing an AC simulation. Reporting on a plot the ratio between the voltage and the current of the imposed external source, the magnitude of the parasitic impedance is derived.

In figure 3.17 it is shown that the parasitic impedance at X node shows a resistive behaviour at lower frequencies and an inductive behaviour for higher frequencies.

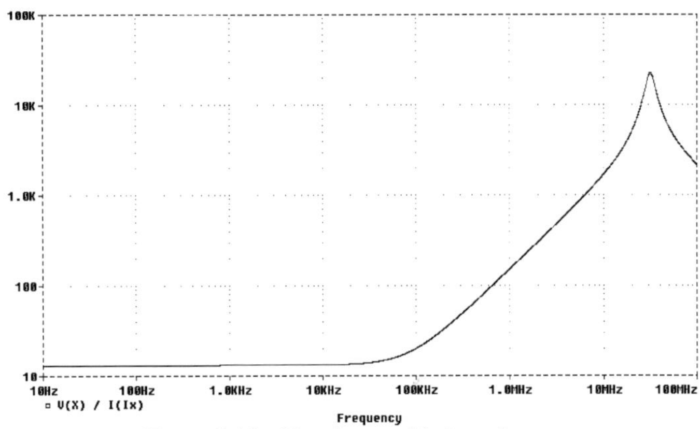

Figure 3.17 – X node parasitic impedance

The impedance at Y node is, as expected, of capacitive type, whose value depends on the input transistor sizes (see figure 3.18).

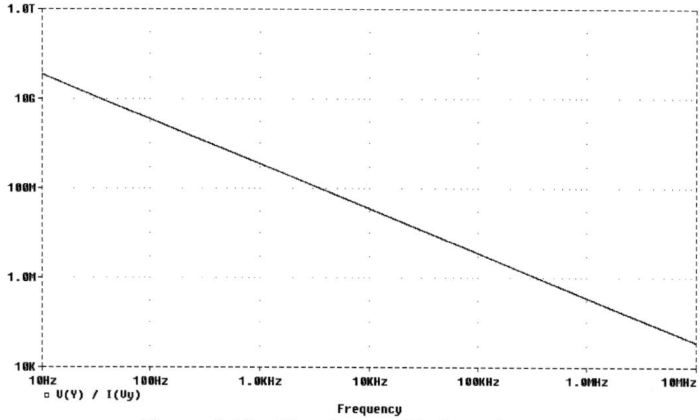

Figure 3.18 – Y node parasitic impedance

The impedance at Z node should be ideally infinite. In real implementations, we can see that a (resistive) high level is obtained, as shown in figure 3.19, while at very high frequencies the capacitive behaviour dominates, so this impedance can be modelled as the parallel of a resistance and a capacitance.

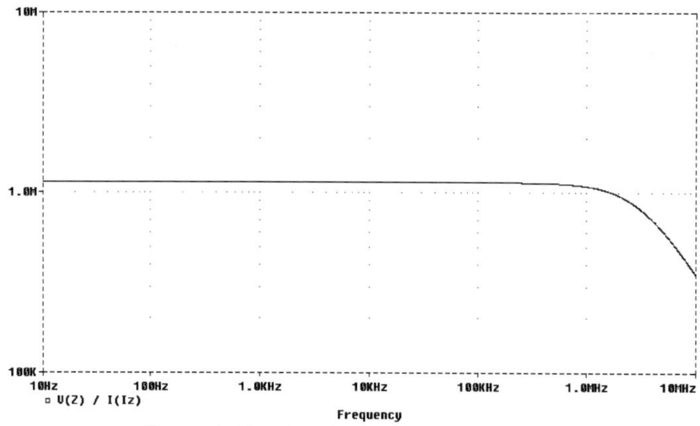

Figure 3.19 – Z node parasitic impedance

The simulations presented in this chapter allow designers to verify the basic CCII performance. Each CCII topology presented in this book has been characterised in this way, and then simulation results have been summarised in a table (see chapter 2 for general CCII structures and the next paragraph for LV LP CCII architectures).

3.3 LOW VOLTAGE LOW POWER CCIIs

CCII-based topologies presented in the previous chapter have been studied at a fixed supply voltage level [9,10,11,12,13,14,15,16,17]. In this paragraph, we are focusing on how is possible to reduce this voltage and which effects this reduction has on CCII performance. This has been done reconsidering all the topologies introduced in the previous chapter at their minimum supply voltage, so evaluating their performance at this supply level, through suitable SPICE simulations. The first current conveyor topology, shown in the figure in table 3.1, has four stacked transistors from positive to negative supply, so preventing a true LV operation [18,19].

Current Conveyor Characteristics

Data	Value
Voltage Supply	± 1 V
Power Consumption	146 μW
3dB Bandwidth	459 MHz
Dynamic Range	– 170 mV, +170 mV
Biasing Current IBIAS1	26 μA
Biasing Current IBIAS2	26 μA
Voltage Gain (α)	0.993
Current Gain (β)	0.98 (Rx=Rz=10KΩ)
Node Y Parasitic Impedance	from KΩ to ∞
Node X Parasitic Resistance	990 Ω
Node Z Parasitic Resistance	282 KΩ

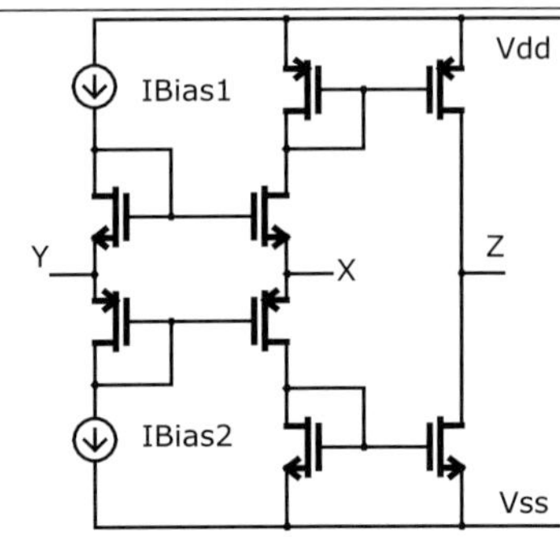

Table 3.1 - LV characteristics for the basic CCII topology

Table 3.1 shows the CCII characteristics, determined at a 2 V total supply. Designing carefully the biasing current sources, it is possible to maintain acceptable small signal characteristics, even if the dynamic range is dramatically reduced.

This behaviour is not surprising, because between X node and the positive supply there must be a minimum voltage drop of:

$$V_{DD}-V_X = V_{TH} + V_{DS} \tag{3.12}$$

so the minimum operating supply voltage is:

$$V_{min} = V_{TH,P} + V_{TH,N} + 2\,V_{DSAT} \tag{3.13}$$

In the considered CMOS technology (AMS 0.35μm), the value indicated in eq. (3.13) is higher than 1.6 V. This explains why the dynamic range is reduced at lower supply voltages. Moreover, even using a higher supply value, the dynamic range is only a smaller fraction of the total supply voltage, so this circuit does not appear suitable for LV use. Moreover, the small signal results reported in table 3.1 show that this circuit, especially for what concerns X node parasitic impedance, needs very high biasing levels, so it has quite poor performance also in terms of LP characteristics.
In order to lower the supply voltage, some class A solutions have been proposed [20]. In table 3.2 the presented circuit has been simulated reducing the total supply voltage to 1.2 V. In order to perform a more reasonable comparison, the biasing current values have been imposed equal to those utilised at higher supply voltages.

Current Conveyor Characteristics	
Data	**Value**
Voltage Supply	± 0.6 V
Power Consumption	18 μW
3dB Bandwidth	15 MHz
Dynamic Range	0 mV, +400 mV
Biasing Current IBIAS1	5 μA
Biasing Current IBIAS2	10 μA
Biasing Current IBIAS3	5 μA
Voltage Gain (α)	0.978
Current Gain (β)	1.18 (Rx=Rz=100KΩ)
Node Y Parasitic Impedance	0.6 pF
Node X Parasitic Resistance	14.7 KΩ
Node Z Parasitic Resistance	3.88 MΩ

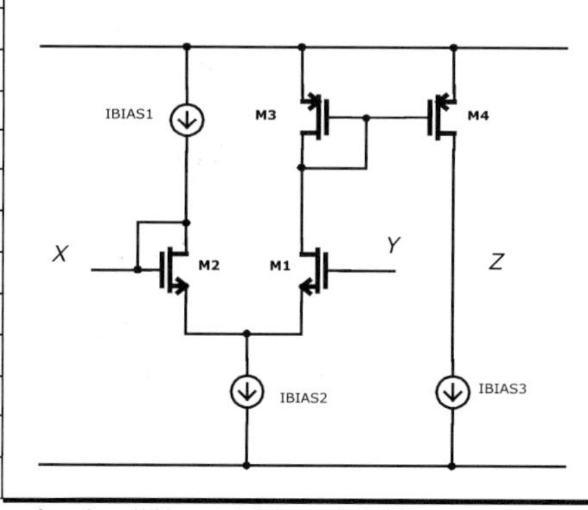

Table 3.2 – LV characteristics for the differential-based CCII

In this case the minimum voltage supply is lower, being equal to:

$$V_{min} = V_{TH} + 2\,V_{DSAT} \tag{3.14}$$

In eq.(3.11), each V_{DSAT} voltage drop takes into account the related biasing source, implemented through current mirrors. Connecting two 10 KΩ resistive loads to X and Z nodes, the dynamic range is practically reduced to zero, so higher loads have to be used to evaluate CCII performance.

This fact represents the greater limit of this solution, even if the other characteristics seem to be acceptable. In conclusion, it possible to say that also this topology, even if shows a lower minimum supply voltage value, can be employed with good performance only at relatively high supplies.
Through the use of a differential pair as CCII input voltage buffer, it is possible, as proved in the previous chapter, to develop a number of CCII topologies. The first solution is presented in figure 3.20, but, owing to its circuit architecture, cannot be taken into consideration for a true LV operation. For this reason, it can be modified towards the circuit proposed in the figure in table 3.3 [21].

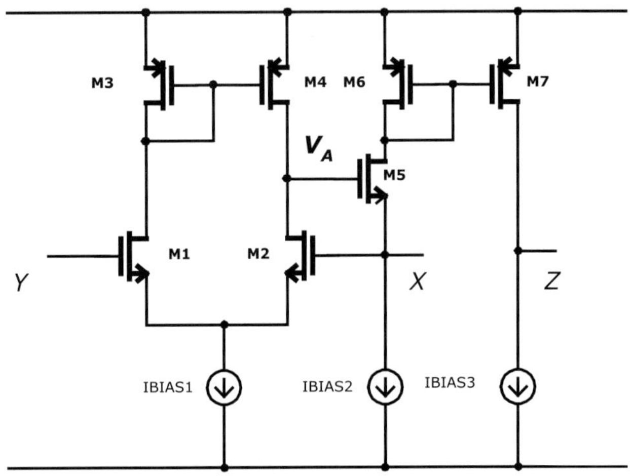

Figure 3.20 – Improved CCII topology based on a differential pair, not suitable for LV operation

In fact, it is possible to obtain a higher impedance level at Y node without the need of very high impedance biasing current sources. Moreover, the class A output stage, designed using only a p-type transistor, increases the output dynamic range, only limited by the fact that output signal variations tend to reduce the biasing characteristics of some transistors, and consequently to switch "off" some of them. This is the great improvement introduced by this solution. With this topology, the output signal may increase up to $V_{DD}-V_{DSAT}$, because the differential pair ensures that M5 transistor is correctly biased. As a result, the minimum supply is still given by eq. (3.11), but the dynamic range is increased.

Table 3.3 presents also the main results, determined at a 1 V total supply voltage, where the CCII characteristics are really very close to those obtained at 1.5 V.

Current Conveyor Characteristics

Data	Value
Voltage Supply	±0.5 V
Power Consumption	30 μW
3dB Bandwidth	21 MHz
Dynamic Range	0 mV, +250 mV
Biasing Current IBIAS1	10 μA
Biasing Current IBIAS2	10 μA
Biasing Current IBIAS3	10 μA
Voltage Gain (α)	1.003
Current Gain (β)	1.00 (Rx=Rz=10KΩ)
Node Y Parasitic Impedance	0.6 pF
Node X Parasitic Resistance	37 Ω
Node X Parasitic Inductance	1.6 mH
Node Z Parasitic Resistance	1 MΩ

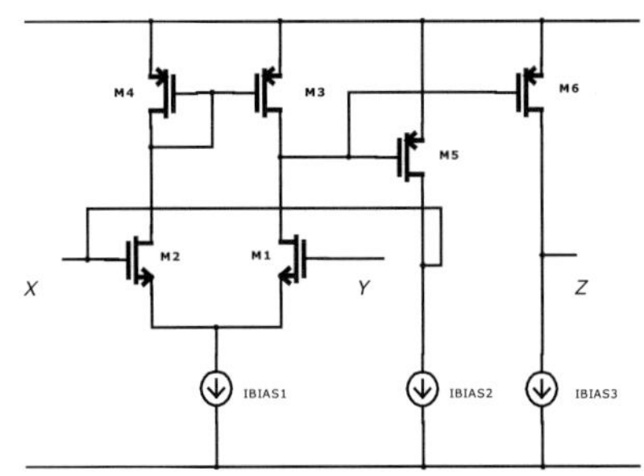

Table 3.3 – LV characteristics of a class A CCII

As stated before, using only one transistor - instead of two - in the X-node output stage, a significant improvement in dynamic range response has been obtained. Moreover, this solution has allowed a lower supply level without affecting in a unacceptable way the overall circuit characteristics.
If the proposed CCII has to drive low valued loads connected to X and Z nodes, there will be a reduction in the operating dynamic range. For example, using two 10 KΩ resistive loads, the signal range is limited to 180 mV.
But the stronger limitation of this solution is its class A operation, which, as well known, is not suitable for really LV applications (in fact, even if the dynamic range at X node has been increased, it is not complete). This solution is also not suitable for LP ones, because, if small loads have to be driven, large biasing currents have to be designed, with consequent high dissipation.
For this reason a class AB topology has been implemented, simply replacing the biasing sources with inverter output stages. This solution is now suitable for LV applications, and the main performance of the CCII implemented in this way are presented in table 3.4 [13,14,15,22]. These characteristics have been determined at 1.2-V total supply, because the output stage cannot work properly at lower values.

Current Conveyor Characteristics

Data	Value
Voltage Supply	± 0.6 V
Power Consumption	36 μW
3dB Bandwidth	16 MHz
Dynamic Range	–450 mV, +450 mV
Biasing Current IBIAS1	10 μA
Current in M5,M6	10 μA
Voltage Gain (α)	1.0007
Current Gain (β)	1.00 (Rx=Rz=10KΩ)
Node Y Parasitic Impedance	0.4 pF
Node X Parasitic Resistance	12 Ω
Node X Parasitic Inductance	430 μH
Node Z Parasitic Resistance	1 MΩ

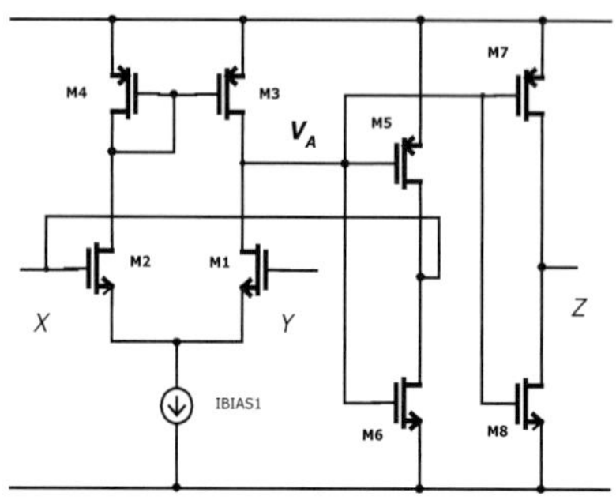

Table 3.4 - LV characteristics of a class AB CCII

The circuit shown in the figure in table 3.5 is another OTA-based CCII [23]. The biasing circuit related to X node prevents a LV operation. With respect to the topology presented in the figure in table 3.1, this solution has the advantage to show a very high input impedance at Y node, given by the gate of a MOS transistor. Using a differential pair, a wider dynamic range has been obtained too, even if the output stage used for X node prevents a real LV operation. Moreover, the feedback connection allows to obtain a lower parasitic impedance at X node. Finally, it must be noted that LV philosophy brings towards different topologies for the X node output stage. Then, CCII characteristics have been determined at a total supply voltage of 2 V.

Current Conveyor Characteristics

Data	Value
Voltage Supply	± 1 V
Power Consumption	67 μW
3dB Bandwidth	10 MHz
Dynamic Range	–500 mV, +150 mV
Biasing Current IBIAS1	3 μA
Biasing Current IBIAS2	10 μA
Voltage Gain (α)	0.993
Current Gain (β)	0.997 (Rx=Rz=10KΩ)
Node Y Parasitic Impedance	0.4 pF
Node X Parasitic Resistance	1.3 Ω
Node X Parasitic Inductance	55 μH
Node Z Parasitic Resistance	2.16 MΩ

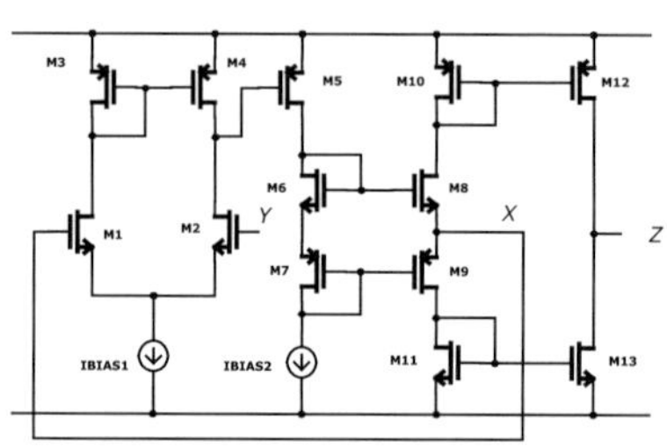

Table 3.5 - LV characteristics of another OTA-based CCII

Let us consider now the circuit in table 3.6, which is able to operate at only 1 V total supply voltage (since that the nMOS threshold voltage, for the AMS 0.35 μm CMOS technology, utilised as the reference in this book, is of about 0.5 V) [24]. This lower operating supply voltage is possible thanks to the development of a different biasing circuit for X and Z output stages, performed through the current generator named IBIAS2 and resistance R. The circuit solution presented for the output stages is particularly suitable for a LV implementation, even if this affects CCII performance, as summarised in table 3.6. Since this CCII is based on a n-type differential pair, the dynamic range is more extended towards the positive voltages. This biasing solution ensures that the p-type transistor M5 is switched "on" even when the voltage applied to Y node is decreased. Anyway, there is a limitation for dynamic range at negative voltages, because X node voltage level cannot follow Y one.

Current Conveyor Characteristics

Data	Value
Voltage Supply	± 0.5 V
Power Consumption	32 μW
3dB Bandwidth	17 MHz
Dynamic Range	–50 mV, +400 mV
Biasing Current IBIAS1	10 μA
Current in M5,M6	10 μA
Current in M7,M8	10 μA
Voltage Gain (α)	1.009
Current Gain (β)	1.00 (Rx=Rz=10KΩ)
Node Y Parasitic Impedance	0.9 pF
Node X Parasitic Resistance	17 Ω
Node X Parasitic Inductance	75 μH
Node Z Parasitic Resistance	925 KΩ

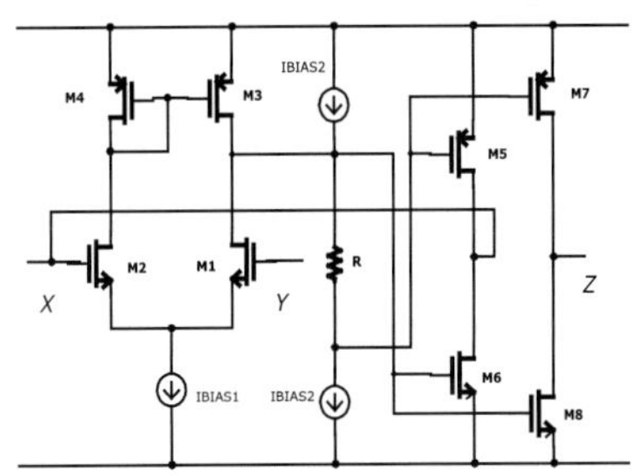

Table 3.6 - LV characteristics of the n-type OTA-based CCII

It is possible to find similar results considering the solution based on the complementary p-type differential pair, whose main results are presented in table 3.7. Moreover, using pMOS transistors, the voltage supply scaling can be limited in a more severe manner. In fact, if the circuit proposed in the figure in table 3.7 is implemented with a standard CMOS technology, the designer must take into account the fact that each single battery used to bias the circuit cannot have a value lower than one threshold voltage (p-type transistors have a threshold voltage of about 0.6 V in AMS 0.35 μm CMOS technology). For this reason, the results summarised in table 3.7 have been determined at a 1.2 V total supply voltage.

Current Conveyor Characteristics	
Data	**Value**
Voltage Supply	± 0.6 V
Power Consumption	36 μW
3dB Bandwidth	16 MHz
Dynamic Range	–600 mV, +200 mV
Biasing Current IBIAS1	10 μA
Current in M5,M6	10 μA
Current in M7,M8	10 μA
Voltage Gain (α)	1.0023
Current Gain (β)	1.00 (Rx=Rz=10KΩ)
Node Y Parasitic Impedance	1.3 pF
Node X Parasitic Resistance	18 Ω
Node X Parasitic Inductance	230 μH
Node Z Parasitic Resistance	1.04 MΩ

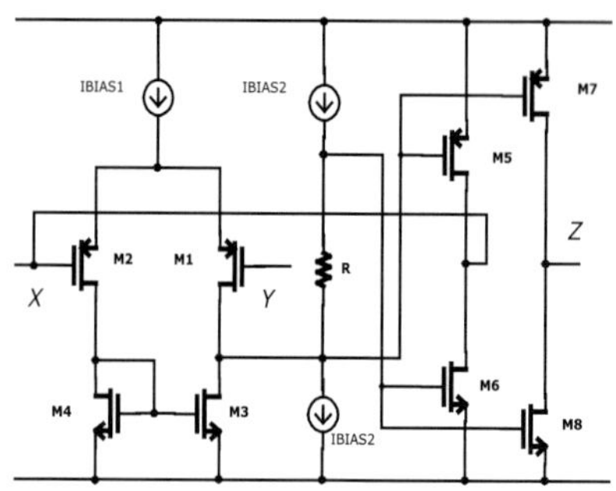

Table 3.7 - LV characteristics of the p-type OTA-based CCII

The two last topologies can be merged together to design a rail-to-rail CCII, as reported in figure in table 3.8. The basic concept is to use both differential pairs, the n-type based and the p-type one, connecting the two single ended outputs to the two gates of the output stages. Formulas presented in chapter 2 show that even if both pairs are active, which can cause g_m problems in common operational amplifiers, the circuit still behaves like a current conveyor. Its performance, summarised in table 3.8, are similar to the one obtained with the previous solutions, while the real advantage is represented by the rail-to rail behaviour. This feature has a particular importance in the LV philosophy, where the reduced voltage range must be utilised in the best possible way. Due to the limitation imposed by the p-type differential pair, the total supply voltage has been set equal to 1.2 V. We have to mention that the presented results have been found without any load connection at X and Z nodes. The dynamic range depends on the load values, even if it is possible, in theory, to design an output stage suitable for any value of load.

For what concerns other possible OTA-based solutions for CCII implementation, they can operate even at a 1 V supply level, as confirmed by eq.(3.14) which is still valid. A possible solution is shown in table 3.9, where its main results in terms of LV CCII characteristics have been also reported.

Current Conveyor Characteristics	
Data	**Value**
Voltage Supply	±0.6 V
Power Consumption	51 μW
3dB Bandwidth	18 MHz
Dynamic Range	–600 mV, +550mV
Biasing Current IBIAS1	10 μA
Biasing Current IBIAS2	10 μA
Current in M5,M6	10 μA
Voltage Gain (α)	1.007
Current Gain (β)	1.000 (Rx=Rz=10KΩ)
Node Y Parasitic Impedance	2.2 pF
Node X Parasitic Resistance	15 Ω
Node X Parasitic Inductance	28 μH
Node Z Parasitic Resistance	1.05 MΩ

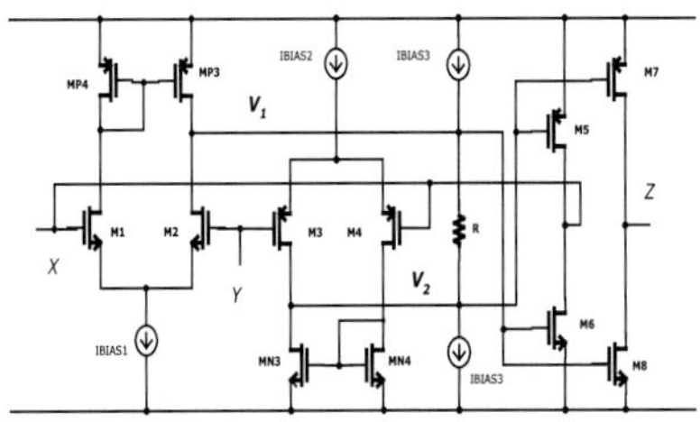

Table 3.8 - LV characteristics of the rail-to-rail OTA-based CCII

Current Conveyor Characteristics	
Data	**Value**
Voltage Supply	±0.5 V
Power Consumption	65 μW
3dB Bandwidth	19 MHz
Dynamic Range	–150 mV, +350 mV
Biasing Current IBIAS1	10 μA
Current in M5,M6	10 μA
Voltage Gain (α)	1.0019
Current Gain (β)	1.000 (Rx=Rz=10KΩ)
Node Y Parasitic Impedance	0.4 pF
Node X Parasitic Resistance	17 Ω
Node X Parasitic Inductance	40 μH
Node Z Parasitic Resistance	951 KΩ

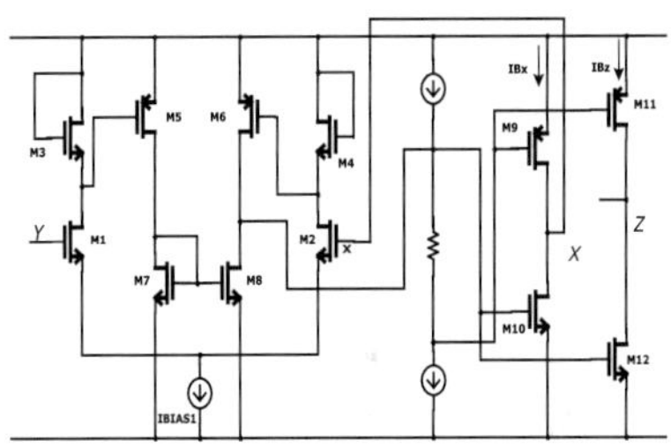

Table 3.9 - LV characteristics of another OTA-based CCII

The use of a symmetrical OTA as differential pair (in its nMOS and pMOS versions, see tables 3.10 and 3.11) allows to obtain good characteristics in terms of input offset and improved dynamic range also at LV supplies.

Current Conveyor Characteristics	
Data	**Value**
Voltage Supply	± 0.5 V
Power Consumption	62 μW
3dB Bandwidth	12.5 MHz
Dynamic Range	–130 mV, +500 mV
Biasing Current IBIAS1	10 μA
Current in M9,M10	10 μA
Current in M11,M12	10 μA
Voltage Gain (α)	1.002
Current Gain (β)	1.00 (Rx=Rz=10KΩ)
Node Y Parasitic Impedance	0.2 pF
Node X Parasitic Resistance	16 Ω
Node X Parasitic Inductance	106 μH
Node Z Parasitic Resistance	950 KΩ

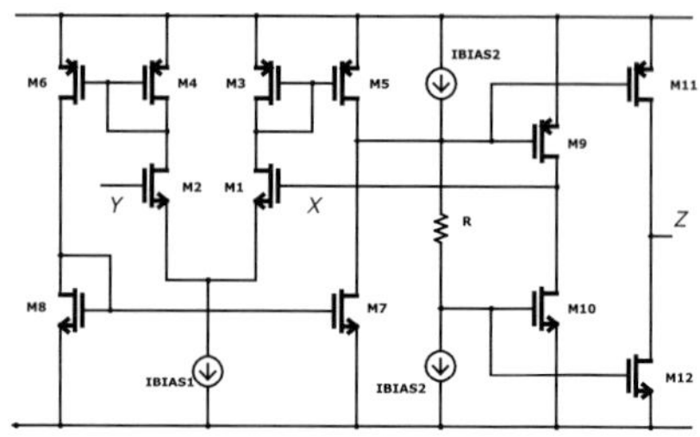

Table 3.10 - LV characteristics of the n-type symmetrical-OTA based CCII

Current Conveyor Characteristics	
Data	**Value**
Voltage Supply	±0.6 V
Power Consumption	70 μW
3dB Bandwidth	12 MHz
Dynamic Range	–600 mV, +400 mV
Biasing Current IBIAS1	10 μA
Current in M9,M10	10 μA
Current in M11,M12	10 μA
Voltage Gain (α)	0.997
Current Gain (β)	1.00 (Rx=Rz=10KΩ)
Node Y Parasitic Impedance	1.06 pF
Node X Parasitic Resistance	17.7 Ω
Node X Parasitic Inductance	125 μH
Node Z Parasitic Resistance	1.07 MΩ

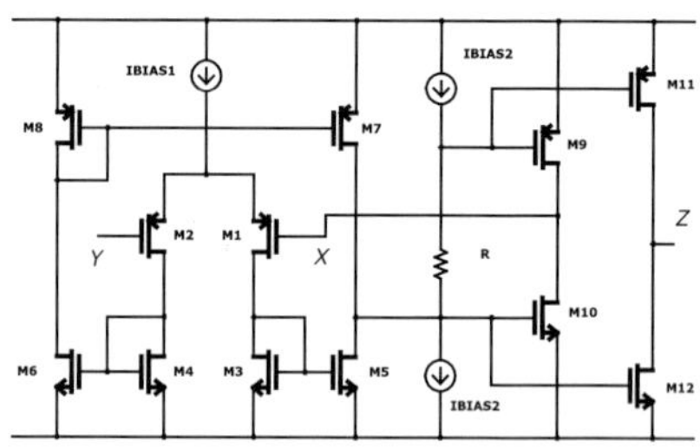

Table 3.11 - LV characteristics of the p-type symmetrical-OTA based CCII

Also in this case, placing the two complementary pairs in parallel, it has been possible to design a rail-to-rail symmetrical OTA based CCII (see table 3.12), which shows very good performance at LV supplies.

Current Conveyor Characteristics			
Data	**Value**	**Data**	**Value**
Voltage Supply	±0.6 V	**Voltage Gain (α)**	1.002
Power Consumption	130 μW	**Current Gain (β)**	1.00 (Rx=Rz=10KΩ)
3dB Bandwidth	13.5 MHz	**Node Y Parasitic Impedance**	0.67 pF
Dynamic Range	–600 mV, +600 mV	**Node X Parasitic Resistance**	14 Ω
Biasing Current IBIAS1	10 μA	**Node X Parasitic Inductance**	78 μH
Current in M9,M10 M11,M12	10 μA	**Node Z Parasitic Resistance**	1.1 MΩ

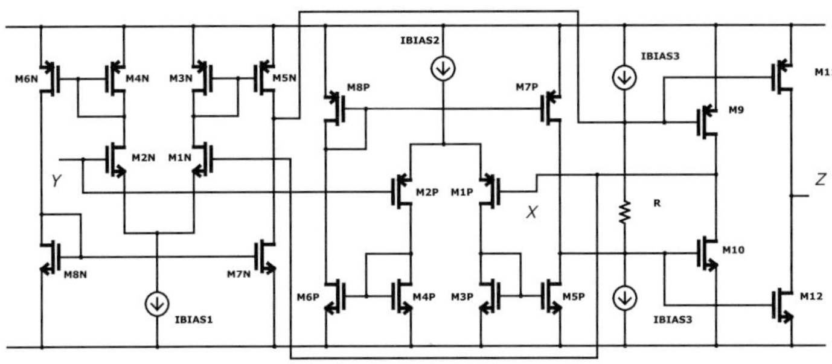

Table 3.12 - LV characteristics of the rail-to-rail symmetrical OTA-based CCII

We can conclude that it is possible to design a number of CCIIs operating at very low supply voltage (± 0.6 V and ± 0.5 V) showing a typical power consumption much lower that 100 μW, with very good performance in terms of bandwidth, voltage and current transfer functions and parasitic characteristics. The circuit shown in table 3.12 presents also the rail-to-rail behaviour, that can be particularly important for low supply voltages, so we believe that this topology can be considered as a good CCII scheme for LV LP applications.

3.4 OFFSET IN CURRENT CONVEYORS

Basically, a CCII can be thought as both a voltage and a current buffer. The voltage buffer has the goal to obtain, at the low impedance X node, a replica of the voltage applied to the high impedance Y node. Obviously, due to the unavoidable non-ideal behaviour of real circuits, there will be an offset between X and Y node voltages, so a further improvement in the CCII model can be considered, as reported in figure 3.21, which considers a positive CCII. An evaluation of the offset is particularly important at reduced supplies, where the signals are lower and all the disturbs have a greater importance.

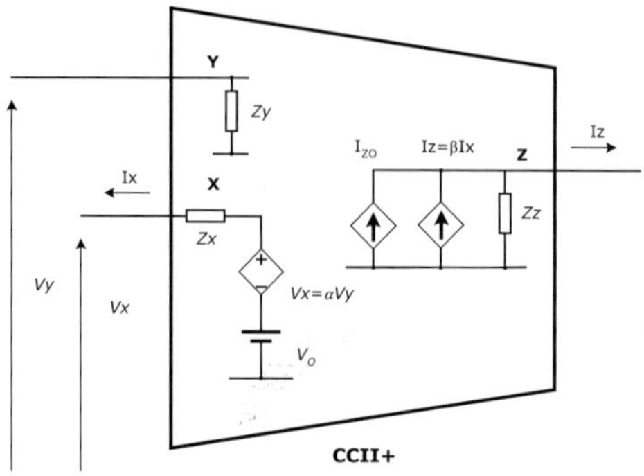

Figure 3.21 – Improved CCII+ equivalent model

The voltage offset V_O causes a current offset I_{ZO}, because of the fact that I_X current is mirrored to high impedance Z node, in order to obtain the output current I_Z. The output current due to voltage offset V_O is dependent on the load connected to X node, in fact:

$$I_X = \frac{\alpha V_Y + V_O}{Z_X + Z_{LOAD}} = \frac{V_O}{Z_X + Z_{LOAD}} \qquad \text{if} \quad V_Y = 0 \tag{3.15}$$

As a consequence, a voltage offset between X and Y nodes may cause a considerable error in the CCII characteristics. In [25], some interesting solutions for offset compensation, in differential pair-based CCIIs, have been proposed. A differential pair should be designed in the most symmetrical way, so to obtain the best performance. In figure 3.22 two schemes of differential pairs are shown.

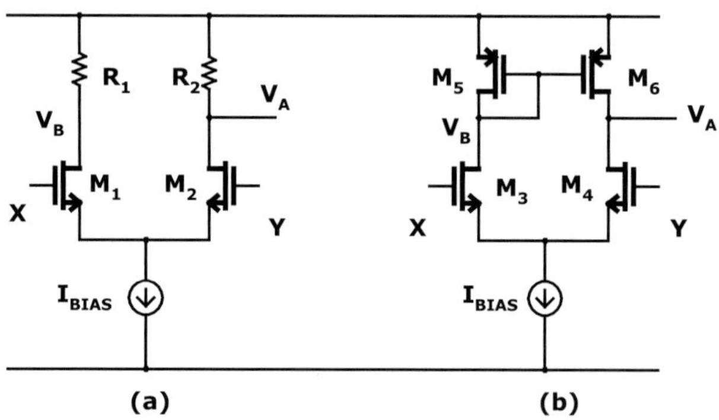

Figure 3.22 – Differential pair schemes: (a) with resistive load – (b) with active load

If the resistive loads connected to M1 and M2 drains are replaced by active loads (so to increase the circuit gain and to reduce chip area), a small difference between V_A and V_B voltages arises, originating a voltage offset between X and Y nodes.
A possible solution can be represented by the circuits proposed in figure 3.23 (a), where V_A and V_B voltages are forced to be equal through the introduction of an external circuit [25].

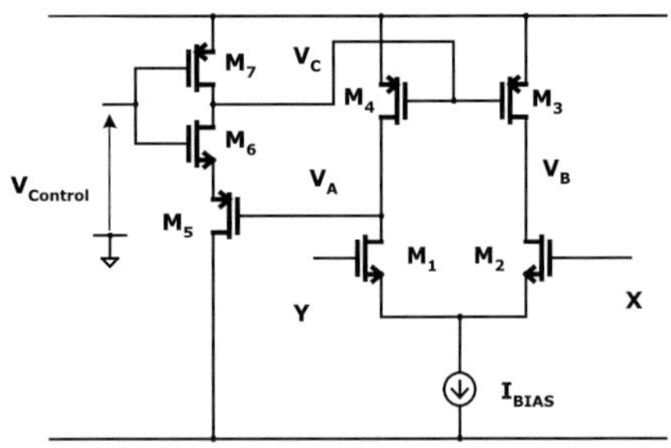

Figure 3.23 (a) –First offset compensation scheme

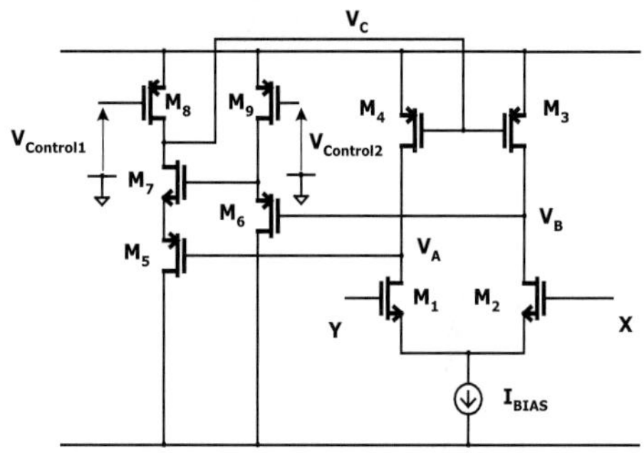

Figure 3.23 (b)–Second offset compensation scheme

Using the solution reported in figure 3.23(a), $V_{Control}$ voltage is trimmed in order to obtain a current conveyor with very low offset. A further improvement is represented by the solution reported in figure 3.23 (b), where the control voltage value is related to the difference between V_A and V_B. Simulations reported in [25] well confirm the theoretical analysis.
The differential pair unbalancing is not the only reason for a voltage offset. Output stage topology and technological spread have to be taken into account too. These two situations have been analysed considering, as example, the CCIIs presented in figures 2.19, 2.21 and 2.23. For a complete analysis, the variations of two parameters have been investigated: the voltage offset at X node and the biasing current of the output stage.

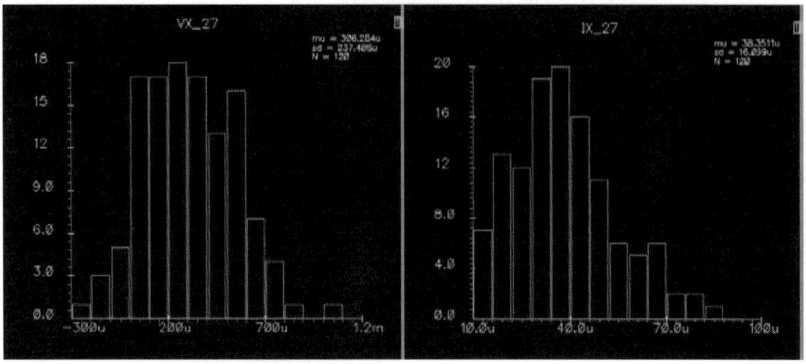

Figure 3.24 – Statistical analysis for figure 2.19 circuit

The output stage utilised in figure 2.19 is strongly affected by voltage supply and technological spread, especially because of the biasing current variations, as reported in the statistical simulations shown in figure 3.24.

The voltage offset has a spread equal to 1.4 mV, while the current spread, considering a biasing current of 30 μA is about 75 μA. In order to overcome these limitations, a different output stage has been introduced, as shown in figure 2.21. In figure 3.25 the statistical analysis for this second circuit is reported. The voltage offset spread is almost the same, while a sensible reduction in the biasing current spread (whose value is now 45 μA) has been obtained. Moreover, it has to be noted also an improvement in the shape of the statistical curves.

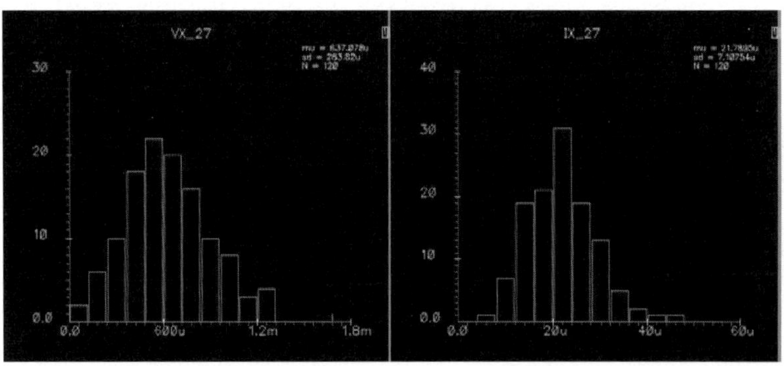

Figure 3.25 – Statistical analysis for figure 2.21 circuit

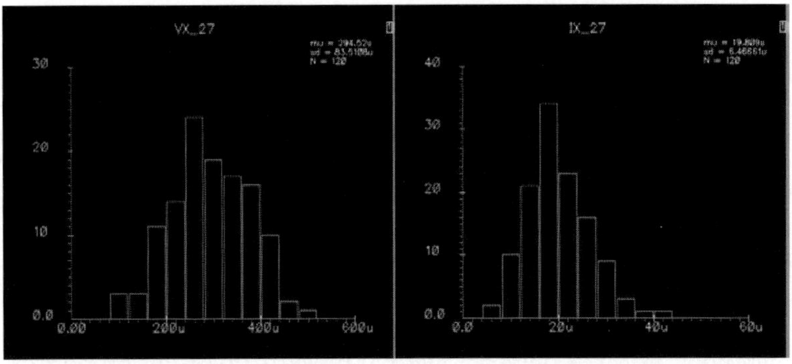

Figure 3.26– Statistical analysis for figure 2.23 circuit

In figure 2.23 a rail-to-rail current conveyor has been presented. The voltage offset performance of this solution are really interesting, because the offset spread has been reduced to only 440 μV. A further improvement for the current spread has been also obtained (35 μA), and also the shape is better.
If current conveyors with good performance are wanted, each detail in the design phase has to be considered. As shown before, a voltage offset between X and Y node may cause considerable errors in CCII characteristics. That's why the differential-pair as well the output stages have to be chosen and designed with care. For what concerns the output stages, it has to be remarked that many other solutions, proposed in literature, can be used.

3.5 NOISE IN CURRENT CONVEYORS

In order to develop good analog circuits, a basic understanding of the noise sources and analysis is required. Noise takes into account the unwanted and unpredictable voltage and current fluctuations in devices and networks.
It is given by two kinds of contributions : external interference (coming from supply, wires, electromagnetic, etc.), which can be reduced by accurate wiring and layout; internal noise (thermal, shot and flicker), due to physical fluctuations inherent to devices, which is more predictable and can be reduced by a proper design of transistor sizes.

3.5.1 Introduction.

In MOS transistors, the main noise contributions are given by thermal and flicker noise. The first is also called "white" noise, because its spectral density is constant over a given frequency. The flicker noise is also known as 1/f noise, because its spectral density is inversely proportional to frequency, so it is dominant at low frequencies. The intersection between flicker and white noise is called 1/f noise corner.

It is known how, in MOS devices, thermal noise is inversely proportional to its transconductance (and, consequently, to its aspect ratio W/L) while flicker noise is inversely proportional to WL product. As a consequence, the choice of transistor sizes is important to minimize noise and has to be done according to the working frequency of the circuit [26].

The thermal motion of electrons is reputed to be the source of what is called "thermal noise". From a classical point of view, a noisy resistor R can be modelled as an ideal (so, noise-free) resistance and an equivalent noise source, as in figure 3.27.

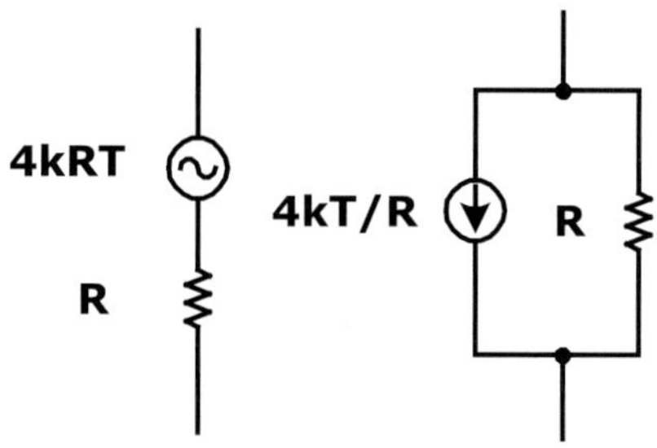

Figure 3.27 – Noisy resistance equivalent models

The noise expressions are given both in terms of voltage and current spectral density, as follows:

$$\overline{v_n^2} = 4kTR \left(\frac{V^2}{Hz}\right) \quad ; \quad \overline{i_n^2} = \frac{4kT}{R} \left(\frac{A^2}{Hz}\right) \tag{3.16}$$

being k the Boltzmann constant ($K=1.38\ 10^{-23}$ J/°K) and T the absolute temperature, expressed in Kelvin degrees.

A MOS transistor can be regarded as a voltage controlled resistance, so its noisy model could be an ideal MOS having in parallel a current noise source dependent on the channel conductance (named g_{DO}).

$$\overline{i_n^2} = 4kT\,\gamma\,g_{DO} \tag{3.17}$$

In eq.(3.17), γ has a value dependent on the operating region of the transistor. Its value is 1 for a transistor in saturation, and 2/3 when its drain-source voltage approaches zero. In the saturation region, this equation becomes:

$$\overline{i_n^2} = 4kT\,\gamma\,g_m \tag{3.18}$$

Obviously, the equivalent noise source can be considered at the input gate terminal, dividing it by g_m^2, as a voltage spectral density.

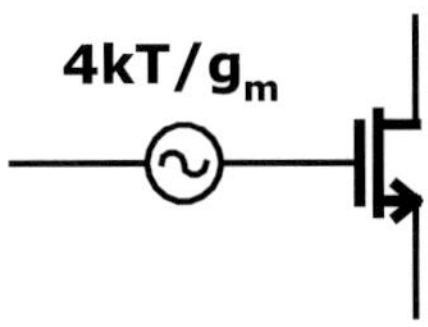

$$\overline{v_n^2} = 4kT\frac{\gamma}{gm} \quad \left(\frac{V^2}{Hz}\right)$$

Figure 3.28 – Noisy model of MOS transistor

Shot noise has been observed especially in all those situations where a current is controlled by a voltage, as, for example, the gate of a MOS transistors. Its formula is:

$$\overline{i_S^2} = 2qI_G \tag{3.19}$$

being q the electronic charge (1.6 10^{-19} C) of the electron and I_G the gate current. As a consequence, in MOS devices, where the gate current is extremely low, shot noise contribution can be easily neglected in a first order evaluation.
Several electronic devices show a noise component strongly affected by frequency. It has been proved, for example, that in some situations noise shows a component inversely proportional to the frequency, that is its level increases when frequency is lowered: this kind of noise has been named “flicker noise” or “1/f noise”. Schottky and Johnson found the flicker noise in thermoionic valves, in 1925.

At the moment, the causes of the flicker noise have not been completely cleared, due to its dependence on the different physical processes involved in transistor operations.
In a MOS transistor, 1/f noise can be modelled by a noise current generator named $i_f^2 = A_f /f$, placed in parallel with the drain-source impedance. Alternatively, it can be taken into account by a series voltage generator, connected to gate terminal, named $v_f^2 = K_f/f$.
K_f and A_f values depend numerically on the technology employed. For example we have [8]:

$$\overline{i_n^2} = \frac{KF\, I_D}{f\, C_{OX} L^2} \left(\frac{A^2}{Hz}\right) \quad \text{and} \quad \overline{v_n^2} = \frac{KF}{2f\, C_{OX} WLK'} \left(\frac{V^2}{Hz}\right) \tag{3.20}$$

where v_n^2 has been obtained from i_n^2 simply dividing by g_m^2. KF (different from K_f) and K' are specific constant, C_{OX} is the unitary gate capacitance and W and L are transistor sizes.
In figure 3.29 the well known small signal model for a MOS transistor is pictured.

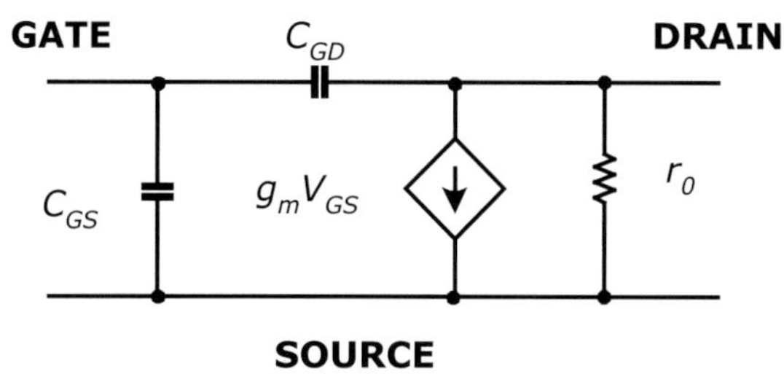

Figure 3.29 –Small signals model of MOS transistor

It is clear that, in this model, noise effects have been not included. This can be done adding the noise generators, previously evaluated, as reported in figure 3.30, where the different noise sources have been modelled as current generators placed at input and output terminals of the MOS transistor. It is also possible to derive an equivalent model of the noisy transistor having all the noise sources at its input, or gate terminal, as in figure 3.31.

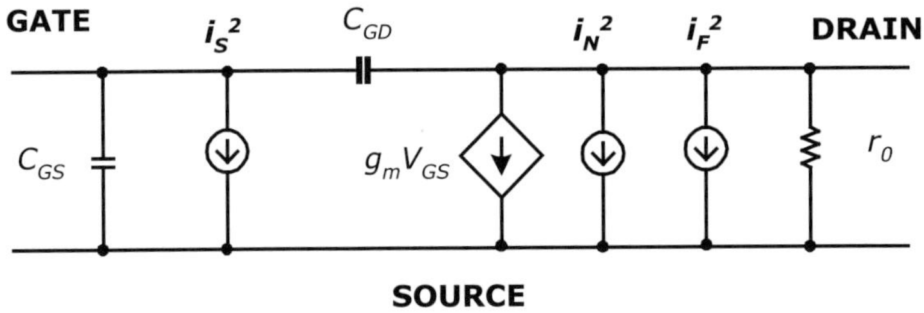

Figure 3.30 –Small signals noisy model of MOS transistor

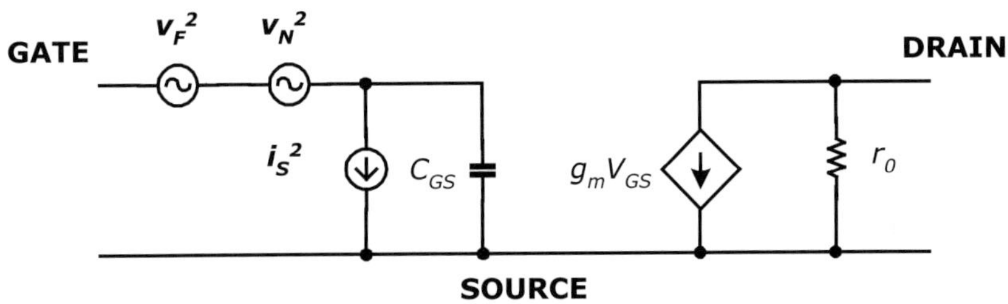

Figure 3.31 –Small signals noisy model of MOS transistor having all noise source at its input

The equivalent input noise voltage sources can be evaluated dividing the output current ones by g_m^2, neglecting the capacitance C_{GD}. This approximation can be done thanks to the fact that 1/f noise is involved in low frequencies operations. Values of γ and, once again, of A_f,, are dependent on the transistor biasing and on the particular technology employed. For MOS transistors it is possible to neglect shot noise component because of the extremely low gate current.
A more simplified model for the "noisy" MOS transistor has only an equivalent input voltage generator $\mathbf{v_T}^2$ or an equivalent output current generator $\mathbf{i_T}^2$, as in figure 3.32. These generators contains the contribution of both thermal and flicker noises.

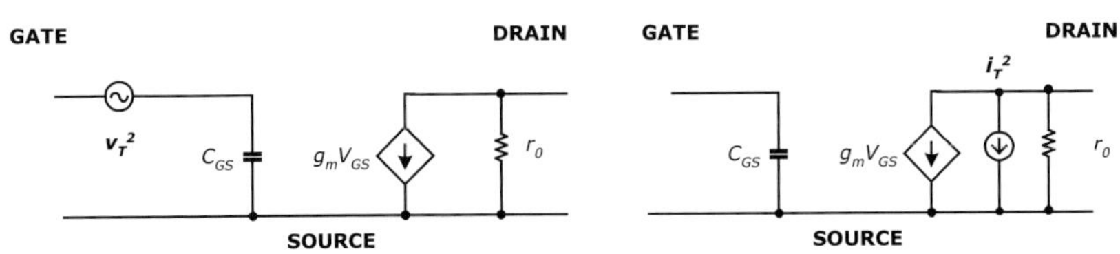

Figure 3.32 - Simplified small signals noisy models of MOS transistor

3.5.2 Noise evaluation in CCIIs.

As for the offset, an evaluation of the noise is more important at reduced supplies, where signals have reduced amplitudes.
The current conveyor topologies presented in figures 2.18 and 2.19 and all their possible improvements follow, basically, the block scheme reported in figure 3.33, where each block has been characterised by its transfer function H(f).

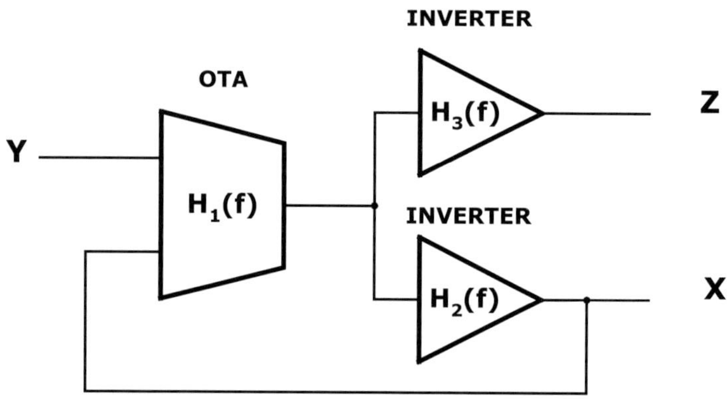

Figure 3.33 – OTA-based CCII block diagram

In fact, from figure 2.19, for example, we have a differential pair, that may be thought as an OTA, followed by two inverting stages. The output of one of these inverters is connected in feedback to one OTA input terminal, so to perform the low impedance X node, while the other inverter is used to obtain the high impedance output node named Z.
Such a kind of scheme allows a different approach to noise analysis, based firstly on the noise determination of each block and then on the analysis of CCII overall noise considering the connections shown in each particular block scheme.
The first block is represented by the input differential pair, reported in figure 3.34 together with the related current noise generators.

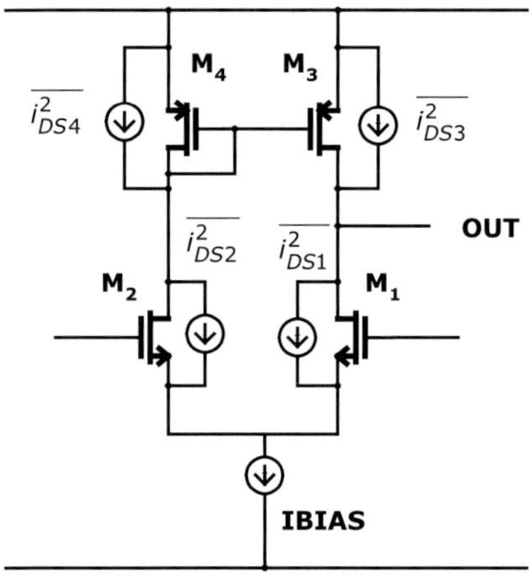

Figure 3.34 – Simple OTA with noise generators

The equivalent output noise current generator, considering all the uncorrelated noise sources, is given by:

$$\overline{i^2_{OUT}} = \overline{i^2_{DS1}} + \overline{i^2_{DS2}} + \overline{i^2_{DS3}} + \overline{i^2_{DS4}} = 4\gamma kT(gm_1 + gm_2 + gm_3 + gm_4) + (A_{f1} + A_{f2} + A_{f3} + A_{f4})\frac{1}{f} \tag{3.21}$$

Moreover, considering the fact that M1–M2 and M3–M
4 can be considered matched, we can write :

$$\overline{i^2_{OUT}} = 4\gamma kT(2gm_2 + 2gm_4) + 2(A_{f2} + A_{f4})\frac{1}{f} \tag{3.22}$$

It is possible to evaluate the equivalent input noise voltage generator, dividing by gm_2^2. Then we have :

$$\overline{v^2_{i1}} = \frac{\overline{i^2_{OUT}}}{gm_2^2} \quad \Rightarrow \quad \overline{v^2_{i1}} = 4\gamma kT\frac{1}{gm_2}\left(2 + 2\frac{gm_4}{gm_2}\right) + 2(A_{f2} + A_{f4})\frac{1}{gm_2^2}\frac{1}{f} \tag{3.23}$$

Finally, from the noise analysis point of view, an equivalent model for the noisy OTA is obtained, as shown in figure 3.35, formed by a noise input generator and a noise-free OTA.

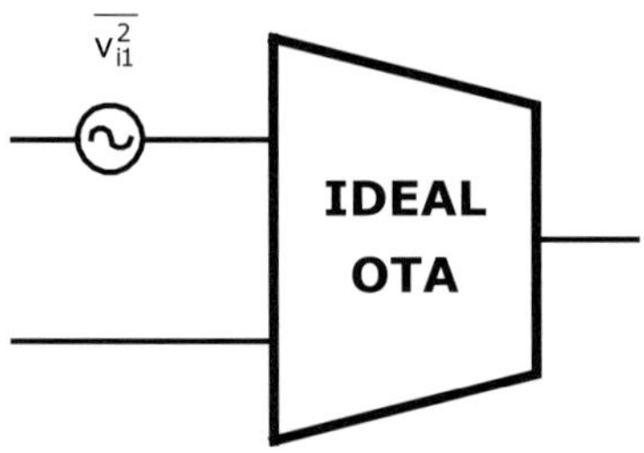

Figure 3.35 – Simple OTA noisy model

The inverter block has to be characterised also from the noise point of view. In figure 3.36, the equivalent input noise generators have been added to the input of the inverter topology.

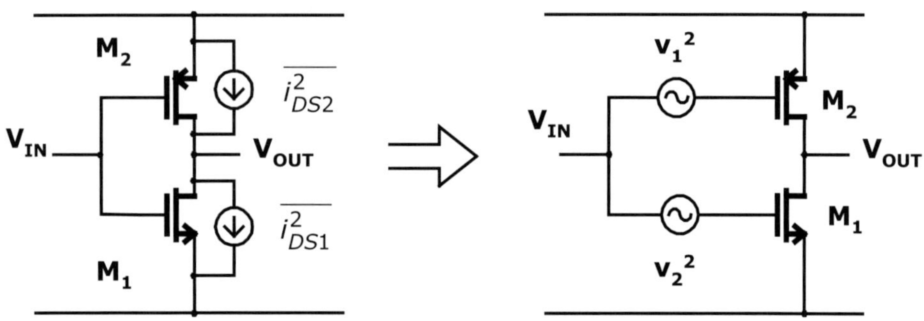

Figure 3.36 – CMOS inverter with noise generators

The small signal equivalent circuit, completed with the noise generators, is reported in figure 3.37, where:

$$\overline{i_{DS1}^2} = 4\gamma kTgm_1 + A_{f1}\frac{1}{f} \quad , \quad \overline{i_{DS2}^2} = 4\gamma kTgm_2 + A_{f2}\frac{1}{f} \tag{3.24}$$

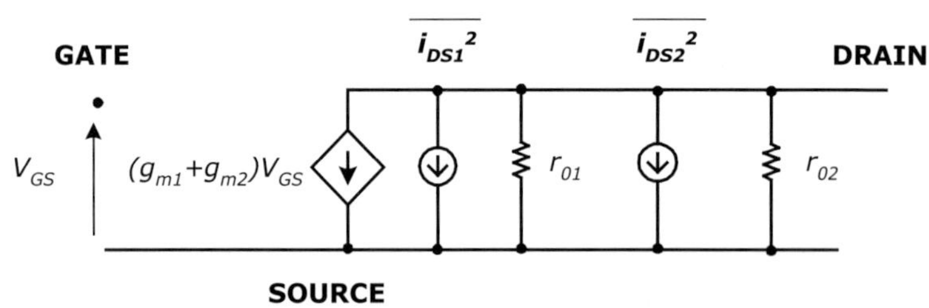

Figure 3.37 – CMOS inverter small signal equivalent circuit with noise generators

The noise generators $\overline{i_{DS1}}^2$ and $\overline{i_{DS2}}^2$ can be easily transferred to the input, as in figure 3.31, so obtaining :

$$\overline{v_{iN}^2} = \overline{v_1}^2 + \overline{v_2}^2 = \frac{\overline{i_{DS1}^2}}{(gm_1 + gm_2)^2} + \frac{\overline{i_{DS2}^2}}{(gm_1 + gm_2)^2} \tag{3.25}$$

As seen for the OTA, an equivalent noisy block for the inverter has been obtained. It is shown in figure 3.38.

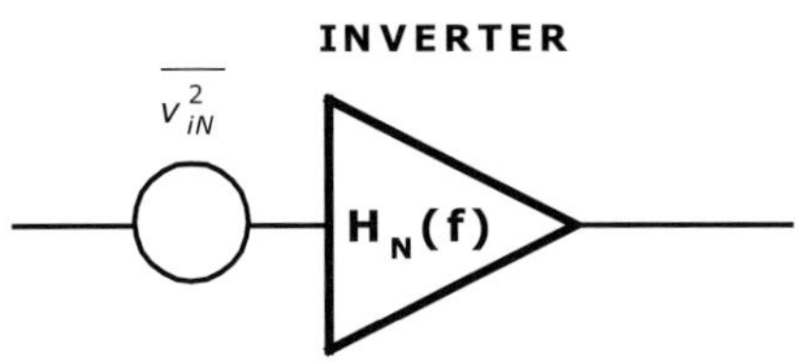

Figure 3.38 – CMOS inverter noise model

Finally, the CCII noisy equivalent circuit can be represented as in figure 3.39, where each stage is considered now noise-free.

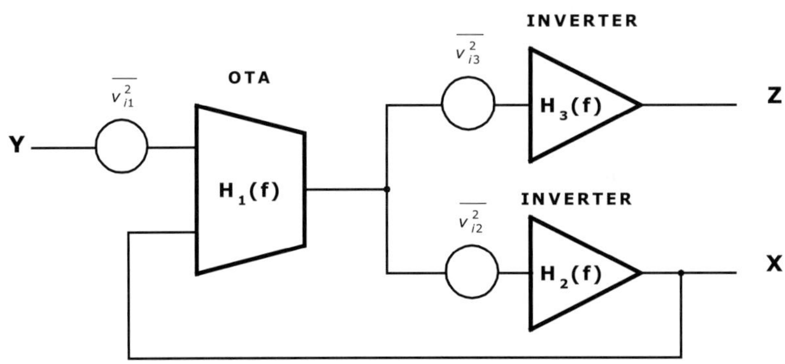

Figure 3.39 – CMOS inverter small signal equivalent circuit with noise generators

For the topology proposed in figure 2.19 it is possible to summarise the noise generators, reported in figure 3.39, as follows:

$$\overline{v_{i1}^2} = 4\gamma kT \frac{1}{gm_2}\left(2 + 2\frac{gm_4}{gm_2}\right) + 2(A_{f2} + A_{f4})\frac{1}{gm_2^2}\frac{1}{f}$$

$$\overline{v_{i2}^2} = \frac{\overline{i_{DS5}^2}}{(gm_5 + gm_6)^2} + \frac{\overline{i_{DS6}^2}}{(gm_5 + gm_6)^2}$$

$$\overline{v_{i3}^2} = \frac{\overline{i_{DS7}^2}}{(gm_7 + gm_8)^2} + \frac{\overline{i_{DS8}^2}}{(gm_7 + gm_8)^2}$$

(3.26)

In order to obtain a useful model for CCII noise and especially to simplify the analysis, all the noise generators have to be referred to Y and X nodes. In the proposed example, we obtain three equivalent input generators (named v_{eq1}, v_{eq2} and v_{eq3}) from the three equivalent noise sources (v_{i1}, v_{i2} and v_{i3}). For what concerns the first noise generator, it is obvious that :

$$\overline{v_{eq1}^2} = \overline{v_{i1}^2}$$

(3.27)

Then, in order to simplify the evaluation of the equivalent Y node noise due to v_{eq2}, related to the first inverter, the CCII block scheme has been modified, as shown in figure 3.40.

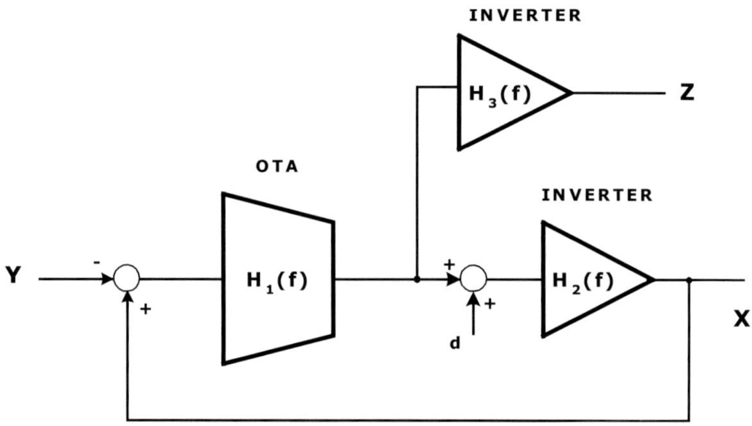

Figure 3.40 – CCII block scheme modified topology

In detail, the OTA has been replaced by a subtracting node and a block having a transfer function equal to $H_1(f)$. The equivalent input noise of the inverter related to X node has been considered as an additive quantity, named **d**.

In this condition, the transfer function between X and Y nodes is:

$$V_X = -\frac{H_1(f)H_2(f)}{1-H_1(f)H_2(f)}V_Y + \frac{H_2(f)}{1-H_1(f)H_2(f)}d \tag{3.28}$$

The noise generator at inverter input can be transferred to CCII input, as in figure 3.41. The equivalent noise at Y node is now evaluated, considering an additive disturb **d'** at Y node and imposing that it gives the same output result.

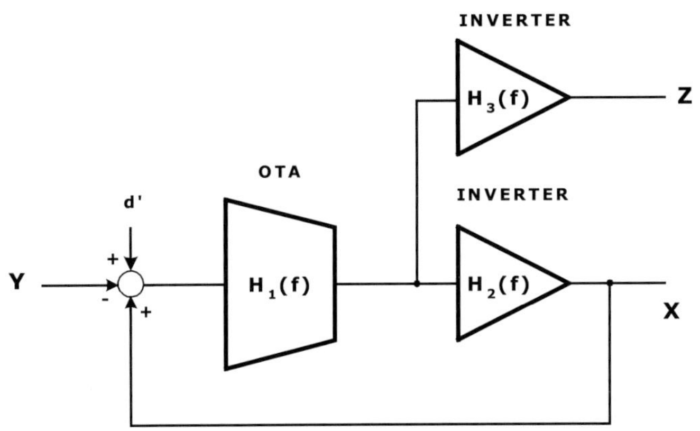

Figure 3.41 – CCII block scheme modified topology, with the disturb at $H_2(f)$ input transferred to Y node

The transfer function between Y and X node is the following :

$$V_X = -\frac{H_1(f)H_2(f)}{1-H_1(f)H_2(f)}V_Y + \frac{H_1(f)H_2(f)}{1-H_1(f)H_2(f)}d' \tag{3.29}$$

The two situations have to be equivalent, so the following relationships have to be verified :

$$H_1(f)d' = d \Rightarrow d' = \frac{d}{H_1(f)} \tag{3.30}$$

$$\sqrt{\overline{v_{eq2}^2}} = \frac{\sqrt{\overline{v_{i1}^2}}}{H_1(f)} \Rightarrow \overline{v_{eq2}^2} = \frac{\overline{v_{i1}^2}}{(H_1(f))^2}$$

(3.31)

A first order approximation allows to write:

$$|H_1(f)| \approx A_{VO1} = gm\frac{rds}{2} >> 1$$

(3.32)

so v_{eq2} can be neglected in the determination of the equivalent input noise at Y node.
To complete this evaluation, the noise v_{i3} has to be considered, and its contribution at Y node v_{eq3} has to be taken into account. This can be done considering the circuit in figure 3.42.

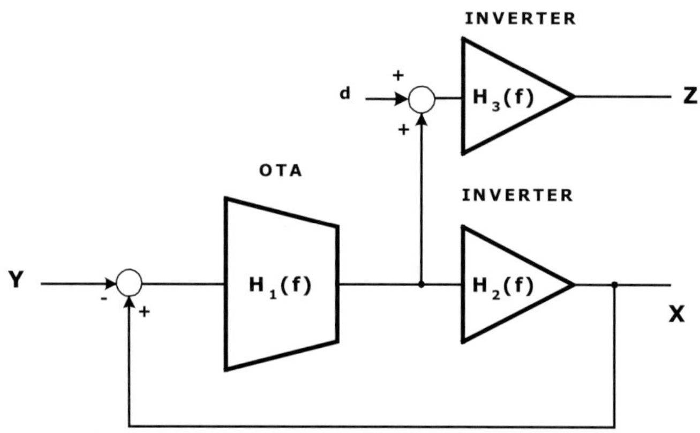

Figure 3.42 – CCII block scheme modified topology

In this circuit, equivalent to the one proposed in figure 3.39, the voltage transfer function between Y and X nodes is:

$$V_X = -\frac{H_1(f)H_2(f)}{1 - H_1(f)H_2(f)} V_Y$$

(3.33)

Following the same approach, the equivalent scheme in figure 3.43 has to be considered.

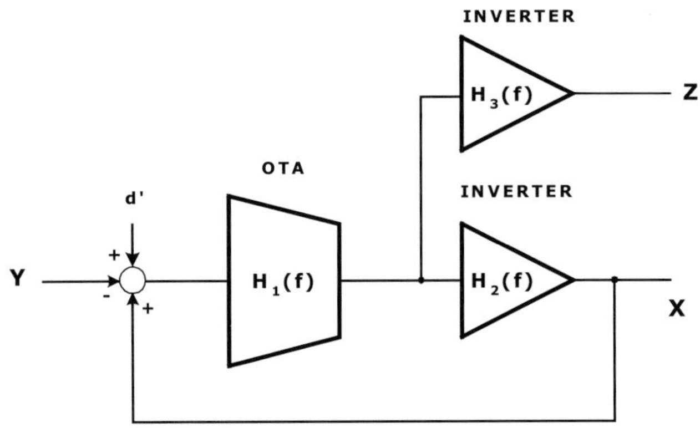

Figure 3.43 – CCII block scheme modified topology, with the disturb at $H_3(f)$ input transferred to Y node

We have :

$$V_X = -\frac{H_1(f)H_2(f)}{1 - H_1(f)H_2(f)} V_Y + \frac{H_1(f)H_2(f)}{1 - H_1(f)H_2(f)} d' \tag{3.34}$$

$$d' = 0 \quad \Rightarrow \quad \overline{v_{eq3}^2} = 0 \tag{3.35}$$

This means that the inverter related to Z node does not give noise contributions at Y node. Then we can write:

$$\overline{v_y^2} = \overline{v_{eq1}^2} + \overline{v_{eq2}^2} + \overline{v_{eq3}^2} \approx \overline{v_{eq1}^2} = 4\gamma kT \frac{1}{gm_2}\left(2 + 2\frac{gm_4}{gm_2}\right) + 2(A_{f2} + A_{f4})\frac{1}{gm_2^2}\frac{1}{f} \tag{3.36}$$

To complete the evaluation of the current conveyor noise sources, the equivalent current noise at X node has to be calculated. The approach is exactly the same used for the Y node noise estimation.

The current conveyor block scheme reported in figure 3.39 has been once more modified, as in figure 3.44, considering Z node as the output of the whole system. The circuit reported in figure 3.44 allows the evaluation of the noise at X terminal due to the OTA, or $H_1(f)$.

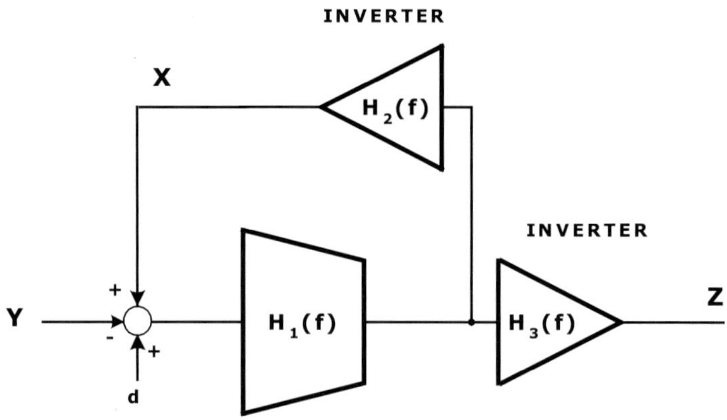

Figure 3.44 – CCII block scheme modified topology

In fact, considering the Y node input noise as an additive disturb **d,** it is possible to write:

$$V_Z = -\frac{H_1(f)H_3(f)}{1 - H_1(f)H_3(f)} V_Y + \frac{H_1(f)H_3(f)}{1 - H_1(f)H_3(f)} d \qquad \text{being:} \qquad d = \sqrt{\overline{v_{i1}^2}} \tag{3.37}$$

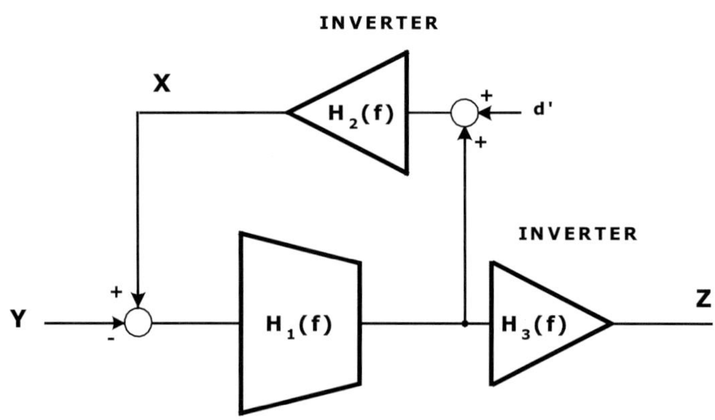

Figure 3.45 – CCII block scheme modified topology with the disturb at $H_1(f)$ input transferred to X node

The equivalent noise at X node is evaluated as before, considering an additive disturb **d'** at X node and imposing that it gives the same output result. For the circuit in figure 3.45 we can write:

$$V_Z = -\frac{H_1(f)H_3(f)}{1-H_1(f)H_3(f)}V_Y + \frac{H_1(f)H_2(f)H_3(f)}{1-H_1(f)H_3(f)}d'$$

(3.38)

from which:

$$d' = \frac{d}{H_2(f)}$$

(3.39)

$$\sqrt{\overline{v_{eq1}^2}} = \frac{\sqrt{\overline{v_{i1}^2}}}{H_2(f)} \Rightarrow \overline{v_{eq1}^2} = \frac{\overline{v_{i1}^2}}{(H_2(f))^2}$$

(3.40)

From the considerations made before to obtain an equivalent noise model for the inverter (figure 3.38), it is possible to say that $H_2(f)$ is quite high in magnitude for low frequencies, so v_{i1} contribution to the overall X node noise can be neglected, in a first order approximation.
Noise contribution v_{i3} at X node has to be considered too. This can be done using the scheme shown in figure 3.46.

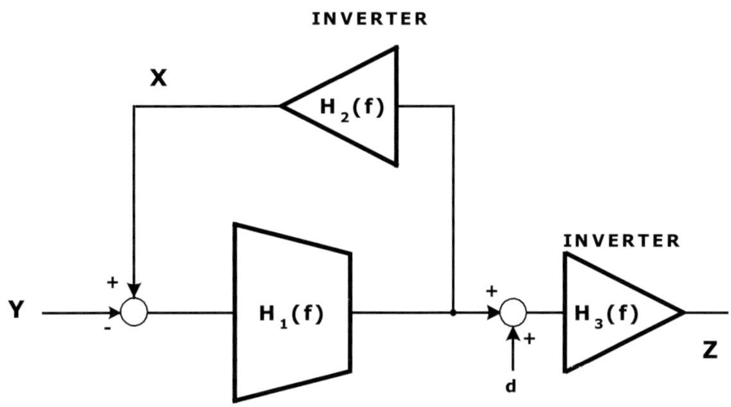

Figure 3.46 – CCII block scheme modified topology considering noise at $H_3(f)$ input

In this scheme the transfer function is given by:

$$V_Z = -\frac{H_1(f)H_3(f)}{1-H_1(f)H_3(f)}V_Y + H_3(f)d \qquad ; \qquad d = \sqrt{v_{i3}^2} \tag{3.41}$$

The equivalent noise at X node is evaluated through the scheme shown in figure 3.47.

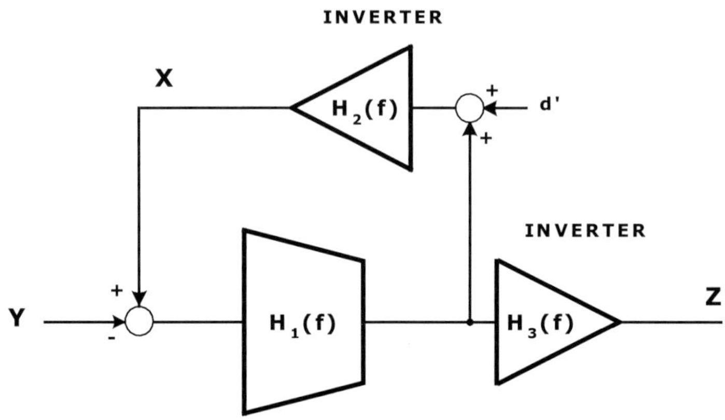

Figure 3.47 – CCII block scheme modified topology with the disturb at $H_3(f)$ input transferred to X node

Following the same approach used in the previous computations, the equivalent noise at X node originated by noise of $H_3(f)$ inverter can be evaluated as follows:

$$d' = \sqrt{v_{eq3}^2} \qquad ; \qquad V_Z = -\frac{H_1(f)H_3(f)}{1-H_1(f)H_3(f)}V_Y + \frac{H_1(f)H_2(f)H_3(f)}{1-H_1(f)H_3(f)}d' \tag{3.42}$$

$$\frac{H_1(f)H_2(f)H_3(f)}{1-H_1(f)H_3(f)}d' = H_3(f)d \;\Rightarrow\; d' = \frac{1-H_1(f)H_2(f)}{H_1(f)H_2(f)}d \tag{3.43}$$

$$\sqrt{\overline{v_{eq3}^2}} = \frac{1 - H_1(f)H_2(f)}{H_1(f)H_2(f)}\sqrt{\overline{v_{i3}^2}} \Rightarrow \overline{v_{eq3}^2} = \left[\frac{1 - H_1(f)H_2(f)}{H_1(f)H_2(f)}\right]^2 \overline{v_{i3}^2}$$

(3.44)

Since we are not interested in a high frequencies noise evaluation, each transfer function $H_i(f)$ in the previous formulas can be approximated by its low frequency value A_{VOi}. Expression for v_{eq3} can be modified as follows:

$$\overline{v_{eq3}^2} \approx \left[\frac{1 - A_{VO1}A_{VO2}}{A_{VO1}A_{VO2}}\right]^2 \overline{v_{i3}^2}$$

(3.45)

$$A_{VO1}A_{VO2} = gm_2 \frac{r_{DS2}}{2} \frac{gm_5 + gm_6}{g_{DS5} + g_{DS6}} >> 1 \Rightarrow \left[\frac{1\text{-}A_{VO1}A_{VO2}}{A_{VO1}A_{VO2}}\right]^2 \approx 1 \Rightarrow \overline{v_{eq3}^2} \approx \overline{v_{i3}^2}$$

(3.46)

Finally, at $H_2(f)$ inverter input, we have three equivalent voltage noise generators, v_{eq1}, v_{eq2} and v_{eq3}.

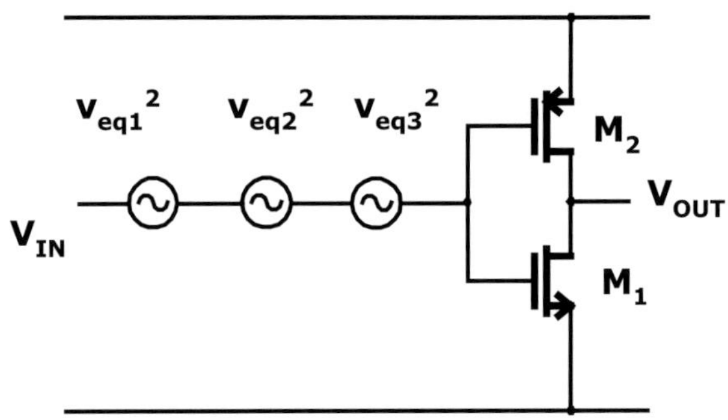

Figure 3.48 – Equivalent noise generators at $H_2(f)$ inverter input

These three voltage generators can be transferred to the output (X node of the CCII) as noise current generators simply multiplying their value by $(g_{m1}+g_{m2})^2$ (considering the figure 3.48) or $(g_{m5}+g_{m6})^2$, considering the current conveyor proposed in figure 2.19. In this last case we can write:

$$\overline{i_x^2} = (gm_5 + gm_6)^2\left(\overline{v_{eq1}^2} + \overline{v_{eq2}^2} + \overline{v_{eq3}^2}\right)$$

(3.47)

Since v_{eq1} can be neglected, and considering the fact that $g_{m5}+g_{m6}$ has to be equal to $g_{m7}+g_{m8}$ to have a good CCII (see eq.(2.35)), we can write :

$$\overline{i_x^2} \approx (gm_5 + gm_6)^2\left(\overline{v_{i2}^2} + \overline{v_{i3}^2}\right) = 4\gamma kT(gm_5 + gm_6 + gm_7 + gm_8) + (A_{f5} + A_{f6} + A_{f7} + A_{f8})\frac{1}{f}$$

(3.48)

Finally, a model for the "noisy" current conveyor can be drawn, as in figure 3.49.

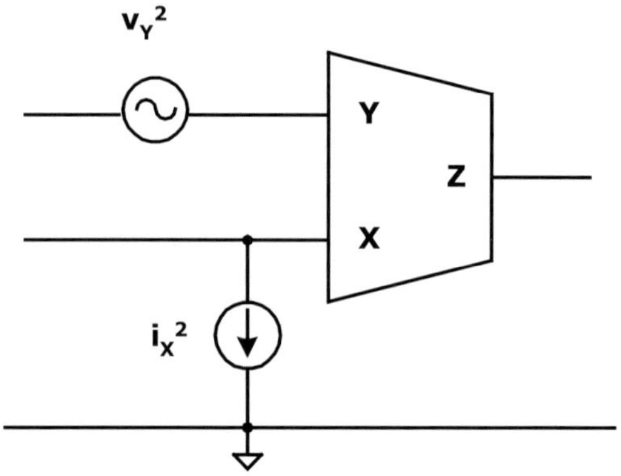

Figure 3.49 – Complete equivalent model for "noisy" current conveyor

The two equivalent noise sources are given by the following expressions:

$$\overline{v_y^2} = 4\gamma kT\frac{1}{gm_2}\left(2 + 2\frac{gm_4}{gm_2}\right) + 2(A_{f2} + A_{f4})\frac{1}{gm_2^2}\frac{1}{f}$$

(3.49)

$$\overline{i_x^2} = 4\gamma kT(gm_5 + gm_6 + gm_7 + gm_8) + (A_{f5} + A_{f6} + A_{f7} + A_{f8})\frac{1}{f}$$

(3.50)

All the theoretical results have been compared with simulation ones, to check the validity of the proposed models for noise sources in current conveyors. From the expressions given above, it is possible to see that noise is directly affected by transistor operating points by means of γ parameter.

Firstly, thermal noise at Y node has been taken into account, and the results of the comparison are pictured in figure 3.50.

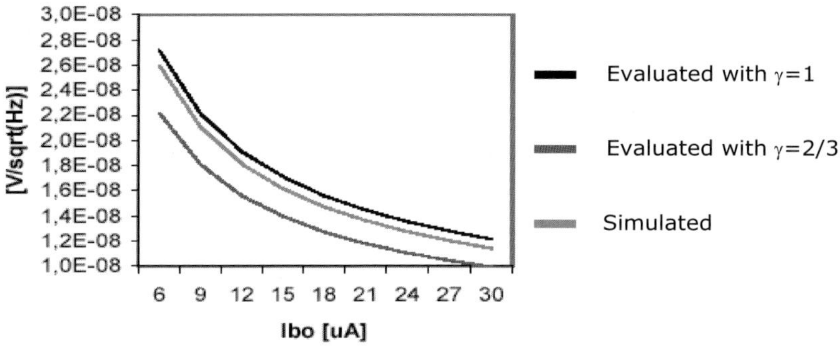

Figure 3.50 – Y node thermal noise vs. differential pair biasing current

The two boundary values for γ parameter have been considered in the theoretical analysis. From figure 3.50 it is clear that the theoretical results are in an excellent agreement with the simulations, being the error between 4 and 5.5%. In order to test if the evaluated noise thermal component at X node is in a good agreement with the practical results, simulations have been performed, modifying the biasing current of the output stages. This can be done varying the W/L ratio. All the stages have been designed having PMOS transistor with an aspect ratio double with respect to the NMOS one. In this way it is possible to consider, to simplify, only a K parameter, that indicates the aspect ratio for NMOS transistors. In figure 3.51 the evaluated and simulated thermal noise component at X node are presented.

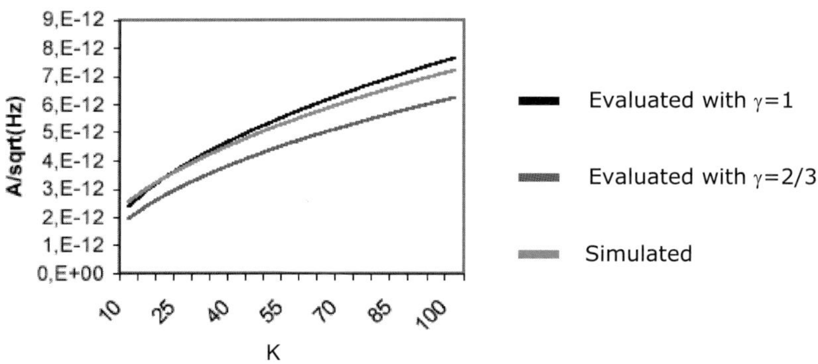

Figure 3.51 – X node thermal noise vs. output stage aspect ratio (biasing current)

Also in this case there is a very good agreement between the results obtained with the proposed noise model and with simulations. The error is always lower than 6%. The same check has to be done on the flicker noise component. For what concerns the voltage noise at Y node, a comparison between theoretical and simulated results is presented in figure 3.52. Once more the difference between them is not high, being the error of about 8%. Flicker noise component has been estimated at X terminal too, and the results are reported in figure 3.53. Also in this case the error is acceptable for a first order evaluation, being lower than 8%.

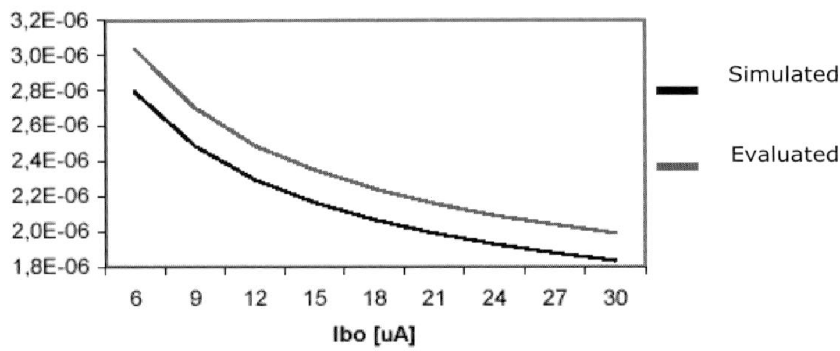

Figure 3.52 – Y node flicker noise vs. differential pair biasing current

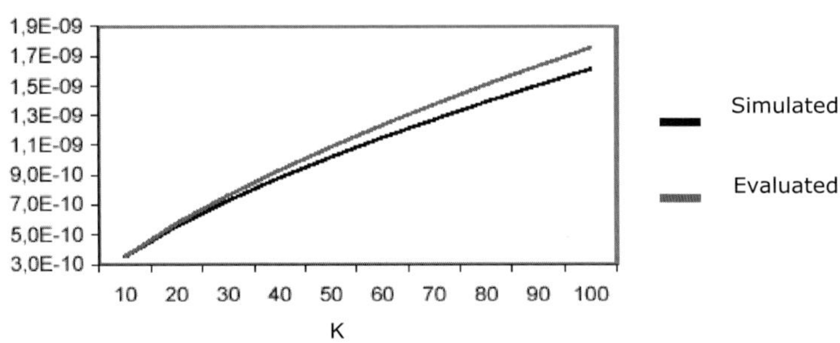

Figure 3.53 – X node flicker noise vs. output stage aspect ratio (biasing current)

In conclusion, it is possible to say that a simple model for the estimation of noise level in current conveyors has been derived. The results obtained shows a little difference with respect to the simulations, but this is due to the fact that we are utilising models and formulas that allow a very fast computation. This is a common trade-off to face. In fact, the more the expressions are detailed, the more they are complex, so requiring a longer time for a results. Anyway, the results of the proposed model can be better detailed if second-order terms, neglected in this derivation, are taken in consideration.

References.

[1] R.Hogervorst, J.Huijsing. *Design of low-voltage low-power operational amplifier cells*. Kluwer Academic Publishers, Boston, 1996.

[2] W.A.Serdijn, A.C. vander Woerd, J.C.Kuenen. *Low-voltage low-power analog integrated circuits*. Kluwer Academic Publishers, Boston, 1995.

[3] S. Sakurai, M.Ismail. *Low-voltage CMOS operational amplifiers*, Kluwer Academic Publishers, Boston, 1995.

[4] G.Ferri. Low power adaptive biased integrated amplifiers. Analog Integrated Circuits and Signal Processing. nr. 33; 2002; pp. 251-264.

[5] G.Ferri, P.De Laurentiis, A.D'Amico, G.Stochino. Low voltage design. Electronics World. September 1999; pp.714-722.

[6] C.Enz, F.Krummenacher, E.Vittoz. An analytical MOS transistor model valid in all regions of operation and dedicated to low-voltage and low-current applications. Analog Integrated Circuits and Signal Processing. nr. 8; 1995; pp. 83-114.

[7] T.Fiez, R.Croman, E.Schnieder, M.Goldeberg. "Current-mode signal processing". In *Analog VLSI : Signal &Information Processing*. Mc-Graw Hill, 1994.

[8] P.E.Allen, D.R.Holberg. *CMOS analog circuit design*, Holt, Rinehart and Winston Inc., New York, 1987.

[9] A. Piovaccari, S. Graffi, G. Masetti. A low voltage low power CMOS current conveyor. Proceedings of the European Conference on Circuit, Theory and Design, 1995.

[10] O. Oliaei, J. Porte. Compound current conveyor (CCII+ and CCII-). Electronics Letters. nr. 4; vol. 33; 1997; pp. 253-254.

[11] A. M. Ismail, A. M. Soliman. Low power CMOS current conveyor. Electronics Letters. nr. 1; vol. 36; 2000; pp. 7-8.

[12] H. O. Elwan, A. M. Soliman. Low voltage low power CMOS current conveyors. IEEE Transactions on Circuit and Systems-I. nr. 9; vol. 44; 1997; pp. 828-835.

[13] G.Ferri, N.Guerrini. High valued passive element simulation using low-voltage low-power current conveyors for fully integrated applications. IEEE Transactions on Circuits and Systems II. nr.4; vol.48; 2001, pp.405-409.

[14] G.Ferri, N.Guerrini. Low-voltage low-power current conveyors : design and applications. Alta Frequenza. nr.4; vol. 12; 2000, pp.59-63.

[15] G.Ferri, N.Guerrini. Low-voltage low-power novel CCII topologies and applications. Proceedings of the IEEE International Conference on Electronic Circuits and Systems, 2001; Malta.

[16] G.Ferri, P.De Laurentiis, G.Stochino. Current conveyors II. Electronics World. April 2001; pp.300-302.

[17] G.Ferri, P.De Laurentiis, G.Stochino. Current conveyors III. Electronics World. May 2001; pp.379-381.

[18] A. Fabre, O. Saaid, H. Barthelemy. On the frequency limitations of the circuits based on second generations current conveyors. Analog Integrated Circuits and Signal Processing. nr. 7; 1996; pp. 113-129.

[19] H.Barthelemy, G.Ferri, N.Guerrini. A 1.5 V CCII-based tunable oscillator for portable industrial applications, Proceedings of the IEEE International Conference on Industrial Electronics, 2002; L'Aquila, Italy.

[20] G. Palumbo, S. Palmisano, S. Pennisi. *CMOS current amplifiers*. Kluwer Academic Publishers, Boston, 1999.

[21] G.Ferri, P.De Laurentiis, G.Stochino. Introduction to current conveyors. Electronics World. February 2001; pp.147-151.

[22] G.Ferri, N.Guerrini, M.C.Piccirilli. Low voltage current conveyor-based universal biquad filter. Proceedings of the IEEE International Conference on Industrial Electronics, 2002; L'Aquila, Italy.

[23] T. Kurashina, S. Ogawa, K. Watanabe. A high performance class AB current conveyor. Proceedings of the IEEE International Conference on Electronic Circuits and Systems, 1998. Lisboa, Portugal.

[24] A. Torralba, private discussion, L'Aquila, July 2002.

[25] I. A. Awad, A. M. Soliman. New CMOS realization of the CCII–. IEEE Transactions on Circuit and Systems-I. nr. 4; vol. 46; 1999; pp. 460-463.

[26] K. R. Laker, W. M. C. Sansen. *Design of integrated circuits and systems*, MC Graw-Hill, New York, 1994.

[27] E. Bruun. Analysis of the noise characteristics of CMOS current conveyors. Analog Integrated Circuits and Signal Processing. nr. 12; 1997; pp. 71-78.

CHAPTER IV

EVOLUTION OF LV LP CCII BASIC BUILDING BLOCK

4.1 IMPROVEMENTS OF THE BASIC CCII

CCII block is powerful and simple at the same time, but the wide spread of possible applications has led to the development of evolutions and improvements of the basic CCII topology.

4.1.1 Dual output CCII (DOCCII).

The simplest modification of the basic CCII topology is represented by its dual output version. As shown in the first chapter of this book, we usually distinguish between positive (CCII+) and negative (CCII-) implementations, depending on the I_Z current sign with respect to the I_X one, assuming the block itself as reference (see figure 4.1).

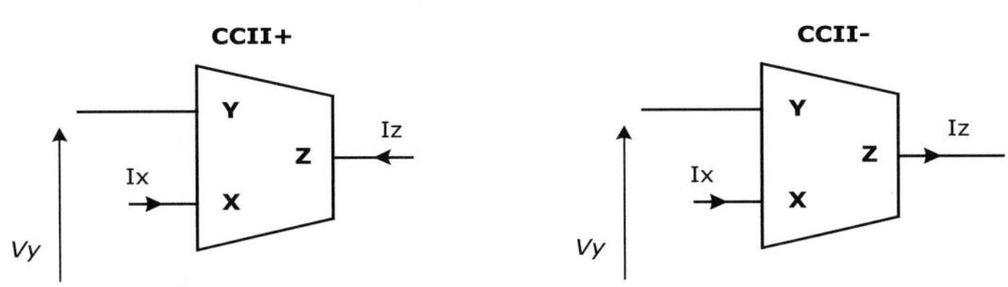

Figure 4.1 – Positive and negative CCII basic blocks

Current conveyors are often employed in applications that require a feedback between input and output terminals. In some applications, it can be useful to have both output currents – inverted and non-inverted – available. For this reason a dual output CCII, whose block scheme is shown in figure 4.2, can be implemented.

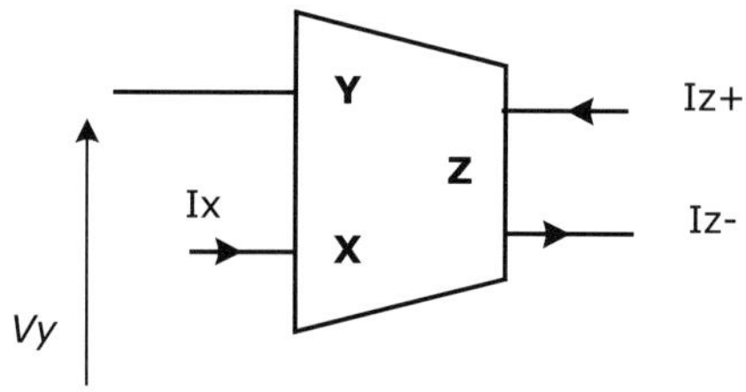

Figure 4.2 – Dual Output CCII basic block

In the previous chapter, several solutions for current conveyor implementation have been presented. A dual output version of a given CCII is easily obtained simply adding some current mirrors at the output node. In this sense, starting, for example, from the circuit shown in figure 2.19, the corresponding dual output version of Z terminal is presented in figure 4.3.

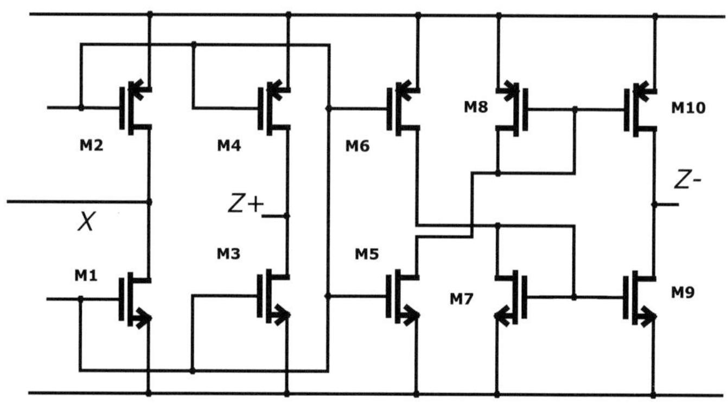

Figure 4.3 – Implementation of a DOCCII output stage

A particular attention in current mirror design has to be paid if better performance of the improved CCII are needed. All the evolutions of the basic CCII towards other building blocks with multiple input and output terminals can be obtained either modifying the internal CCII topology or through suitable connections of more CCII basic blocks and passive components. In this sense, figure 4.4 shows another possible implementation of the dual output CCII.

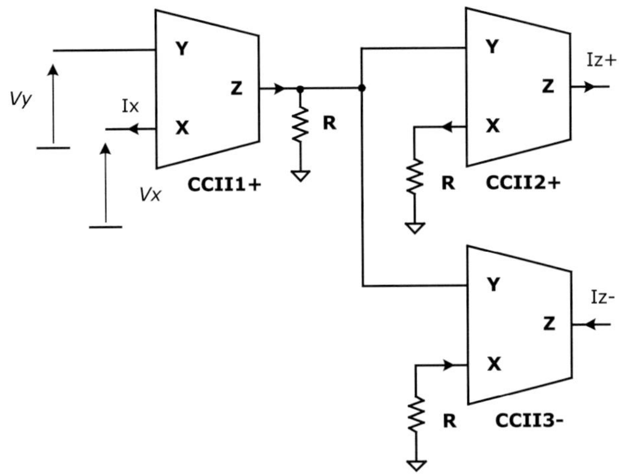

Figure 4.4 –CCII-based Dual Output CCII implementation

The two output currents are obtained through a double conversion. Firstly, the Z node current of CCII1 is forced into a resistance R and then converted into a voltage, RI_Z. Secondly, this voltage is applied to both the Y nodes of CCII2 and CCII3. The current flowing from their Z and X nodes is:

$$I_{Z2,3} = I_{X2,3} = \frac{V_{X2,3}}{R} = \frac{V_{Y2,3}}{R} = \frac{V_{Z1}}{R} = \frac{R\, I_{Z1}}{R} = I_{X1} \tag{4.1}$$

In this manner, two currents equal to that flowing from X1 node have been obtained. They satisfies the requirements for a DOCCII thanks to the fact that CCII3 is a negative current conveyor. This means that $I_{Z2}=-I_{Z3}$ or $I_{Z+}=-I_{Z-}$ (see figure 4.4). It has to be noted that this is true only if the three resistances employed are perfectly matched, otherwise a current transfer error between X and Z nodes will be introduced. The DOCCII can be considered as a simple generalization of the CCII, because adding a current mirror is very simple. Anyway, it helps to demonstrate that all the "evolutions" of the basic current conveyor can be implemented starting from the basic block. In the following pages this principle will be applied to all the other topologies presented before.

4.1.2 Current Gain CCII (CGCCII).

A basic characteristic of current conveyors is represented by the fact that currents flowing at X and Z nodes are almost equal. β parameter has been introduced to quantify the current transfer error between these two terminals, typical of non ideal CCIIs [1,2,3,4,5].
If I_Z is designed K times higher with respect to I_X, the new block so implemented is named Current Gain Current Conveyor (CGCCII) and is shown in figure 4.5.

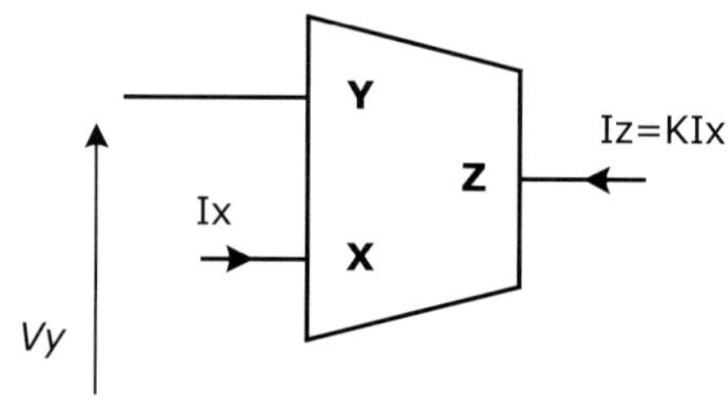

Figure 4.5 – Current Gain CCII basic block

This block has been successfully employed in some applications like capacitance multiplication or inductance simulation (see next chapter). Starting from the basic block, a CGCCII may be designed simply adding some current mirrors, which perform the desired current gain. In figure 4.6 a class A version of a CGCCII output stage is reported.

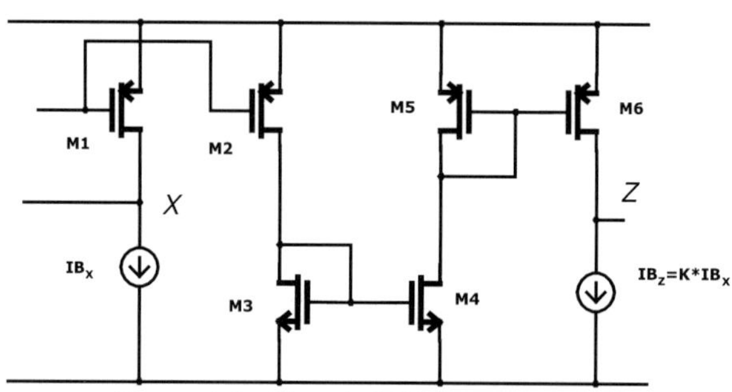

Figure 4.6 – Class A CGCCII output stage

The same principle may be applied to obtain a class AB CGCCII, as that proposed in figure 4.7. Even if the design is quite simple, a particular attention has to be paid to the impedance levels, for which some trade-off considerations have to be done. In fact, imposing a current gain equal to K, the biasing current of Z output stage is increased by the same factor. But a higher current at Z node means a lower impedance at the same terminal, which is typically characterised by a very high impedance level.
The problem could be solved imposing a very low biasing current at X node, but this will cause an increase in X node impedance that, on the contrary, has to be as low as possible.

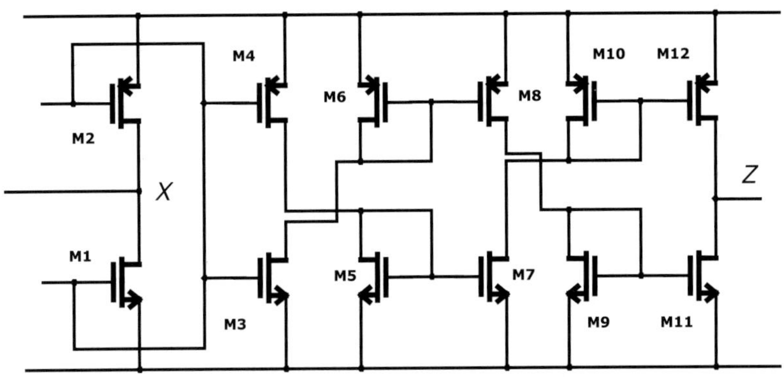

Figure 4.7 – Class AB CGCCII

A different and more efficient approach leads to the use of very high impedance current mirrors (as the "cascoded" ones), which show very high output impedances without particular constraints on biasing current. Anyway, this solution has some limitations in the LV operation, so specific current mirror topologies have to be developed (see chapter 3). This implementation has to used with care. In fact, increasing the output current at Z node by means of current mirrors will lead, in the case reported in figure 4.7, to a lower impedance level.
Using two basic CCIIs and two resistances it is also possible to implement a CGCCII, as reported in figure 4.8.

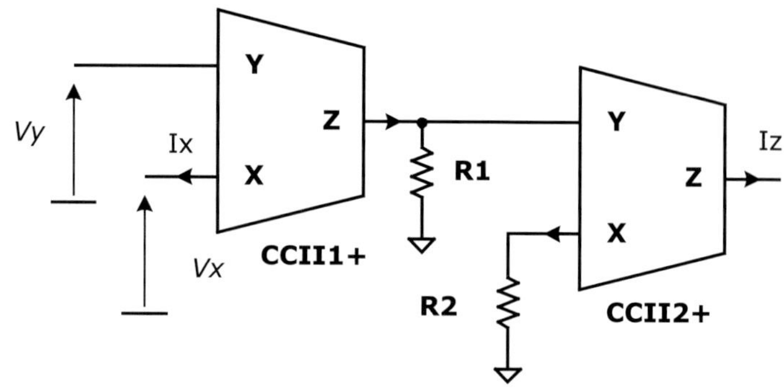

Figure 4.8 – CCII-based current gain CCII implementation

Once more the output current is obtained through a double conversion, but in this case R_1 and R_2 are not equal, so to have a current gain:

$$I_Z = I_{Z2} = I_{X2} = \frac{V_{X2}}{R_2} = \frac{V_{Y2}}{R_2} = \frac{V_{Z1}}{R_2} = \frac{R_1 I_{Z1}}{R_2} = \frac{R_1 I_{X1}}{R_2} \tag{4.2}$$

The limit in the gain that can be imposed is represented by the fact that R_1 can not be too high, or comparable with the parasitic impedance R_{Z1} and R_2 can not be too low, or comparable with the parasitic impedance R_{X2}. If these constraints are not met, an unacceptable difference between the nominal gain (given by the ratio of R_1 and R_2) and the effective gain will result.

4.1.3 Current Controlled CCII (CCCII).

One of the non-idealities of the CCII arises from the fact that the impedance level at X node is low but not zero.

In chapter 2, some formulas expressing the non-ideal impedance at X node, have been proposed for the different topologies of current conveyors presented. These values are affected by the biasing condition of the transistors that form the output stage. Starting from these considerations, the designer can adjust the value of the X node parasitic impedance implementing different topologies and imposing different biasing conditions. Usually, it is necessary to perform a trade-off between a lower parasitic impedance and other specifications, such as power consumption, linearity, dynamic range, and so on.

In some applications, the resistive load connected to X node, which usually represents a limitation for CCII, can give us an advantage. This happens in the device that is named Current Controlled Current Conveyor, or CCCII [6,7].
In this new block, the impedance seen at X node is not more a parasitic element, but becomes a part of the block specifications, as presented in the matrix form reported in figure 4.9, together with CCCII block symbol.

$$\begin{bmatrix} Iy \\ Vx \\ Iz \end{bmatrix} = \begin{bmatrix} 0 & 0 & 0 \\ 1 & R_X & 0 \\ 0 & \pm 1 & 0 \end{bmatrix} \begin{bmatrix} Vy \\ Ix \\ Vz \end{bmatrix}$$

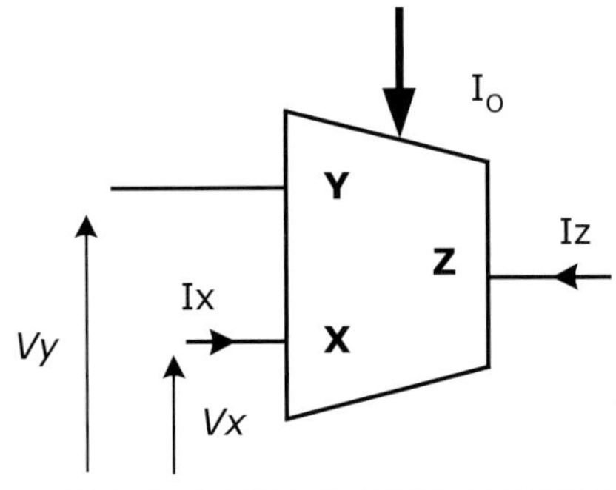

Figure 4.9 – Current Controlled CCII basic block and characteristics

In order to control the biasing of the output stage, the topology presented in figure 2.14 may be modified, for example, in that shown in figure 4.10.

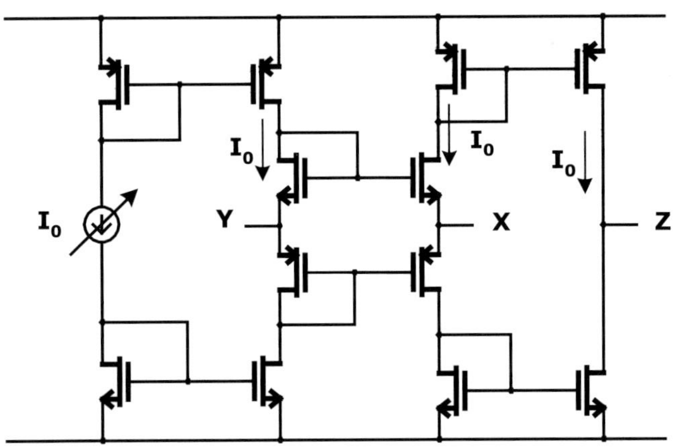

Figure 4.10 – Class AB CCCCII

The current I_0 may control the biasing of the output stage, so modifying the parasitic resistance R_X. In figure 4.11, a graph, showing how R_X changes with the biasing current, is reported. The determined values refer to a simulation in a standard CMOS technology.

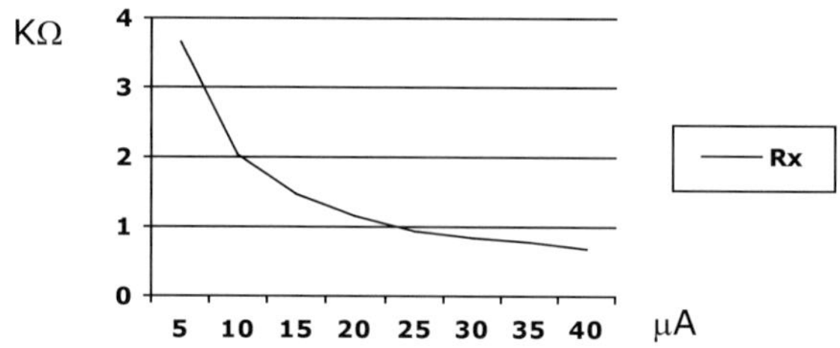

Figure 4.11 – R_X resistance vs biasing current for CCCII

4.1.4 Third generation CCII (CCIII).

Since their introduction, first and second generation current conveyors have been successfully employed in an impressive number of analog applications. It is quite common to express CCI and CCII characteristics in a matrix form. From a general point of view it is possible to consider the two blocks as particular cases of a more general structure [8], described by the following matrix relation:

$$\begin{bmatrix} Iy \\ Vx \\ Iz \end{bmatrix} = \begin{bmatrix} 0 & a & 0 \\ 1 & 0 & 0 \\ 0 & b & 0 \end{bmatrix} \begin{bmatrix} Vy \\ Ix \\ Vz \end{bmatrix} \tag{4.3}$$

The parameter indicated as “b” is the current transfer characteristic. When it is positive we have a CCI+ or a CCII+, while CCI– and CCII– are obtained if $b<0$. A CGCCII is built up when the module of b is greater than one. In (4.1) another parameter, indicated as “a”, has been inserted. If $a = 1$ we have a CCI, while a CCII is obtained when $a=0$. In [8] the case of $a = -1$ has been investigated, leading to the introduction of a new block, named third generation current conveyor or CCIII, whose symbol is reported in figure 4.12.

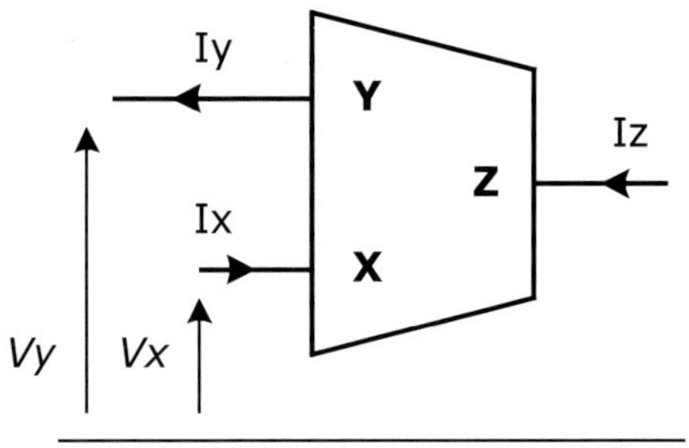

Figure 4.12 – Third generation CCII (CCIII) basic block

Third generation current conveyors may be useful, for example, in current sensing applications [8]. In fact, if a current, at a generic point into a network, has to be sensed, the current "probe" should be able to make flow a current with a very low series impedance and a high impedance current output. This is what a CCIII exactly does, as confirmed by (4.1) if a= –1.
In figure 4.13, a CCII-based CCIII implementation is presented [8].

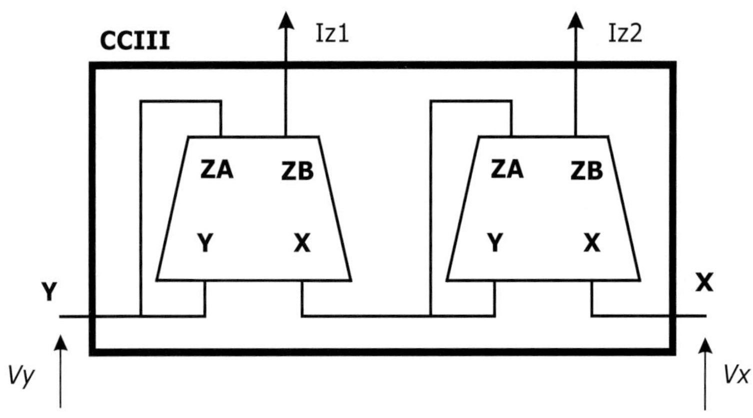

Figure 4.13– CCII-based CCIII implementation

A second output node has been added to the CCIIs employed but, with respect to the philosophy of DO-CCII, we have two equal Z nodes (both named ZB) instead of complementary ones. One of these two ZB terminals can be considered as Z node of the CCIII.

In [8], really good performance for the proposed block have been presented. Implementing a rail-to-rail current conveyor like that introduced in chapter two (figure 2.23), a LV LP CMOS third generation current conveyor can be designed. In table 4.1 typical main CCIII characteristics are summarised.

Current Conveyor Characteristics	
Data	**Simulated Value**
Voltage Supply	±0.75 V
Power Consumption	98 μW
3dB Bandwidth	40 MHz
Dynamic Range	-450mV, + 450mV
Voltage Gain (α)	1.0002
Current Gain (b=β)	1.00 (Rx=10KΩ)
Current Gain (a)	1.00 (Rx=Rz=10KΩ)

Table 4.1 – Third generation current conveyor (CCIII) main characteristics

4.2 TOWARDS THE DIFFERENTIAL SOLUTIONS

LV LP design philosophy is often joined to the need of manipulating differential signals. Designing circuits suitable for differential signals leads to have more versatile applications. In analog design, CCII represents one of the most useful building block. Many efficient applications can be designed with success using CCII as basic component. Anyway, second generation current conveyors, as they have been proposed, show some drawbacks. For example, only one of the input terminals presents a high impedance level. This can be a problem if differential signals have to be handled. To overcome this, a solution using more CCIIs has been proposed [9,10,11,12].
A different approach can be that to implement more complicated basic blocks, which will be presented in this chapter. Each of them can be designed from simple modifications of the basic CCII, so confirming an intrinsic characteristic of the CCII to be "the" basic analog block.

4.2.1 Differential CCII (DCCII).

The search for a new powerful block, in the differential approach, leads to several circuit solutions that propose themselves as natural evolutions of the basic CCII topology.

A first example may be represented by the differential current conveyor (DCCII), characterised by the block scheme and matrix form of figure 4.14 [10].

$$\begin{bmatrix} V_{X1} \\ V_{X2} \\ I_{Z1} \\ I_{Z2} \\ I_{Y} \end{bmatrix} = \begin{bmatrix} 0 & 0 & 0 & 0 & 1 \\ 0 & 0 & 0 & 0 & 1 \\ 1 & -1 & 0 & 0 & 0 \\ -1 & 1 & 0 & 0 & 0 \\ 0 & 0 & 0 & 0 & 0 \end{bmatrix} \begin{bmatrix} I_{X1} \\ I_{X2} \\ V_{Z1} \\ V_{Z2} \\ V_{Y} \end{bmatrix}$$

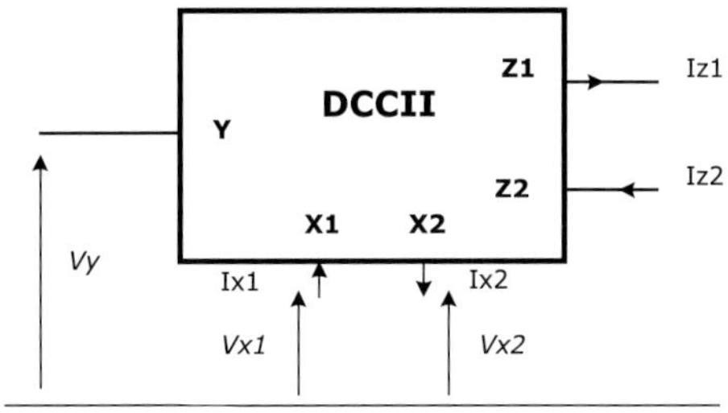

Figure 4.14 –DCCII block representation and characteristics

Both Y and Z nodes are high impedance terminals, while the two X nodes show low impedance levels. In figure 4.15 a class A implementation of DCCII is shown [10].

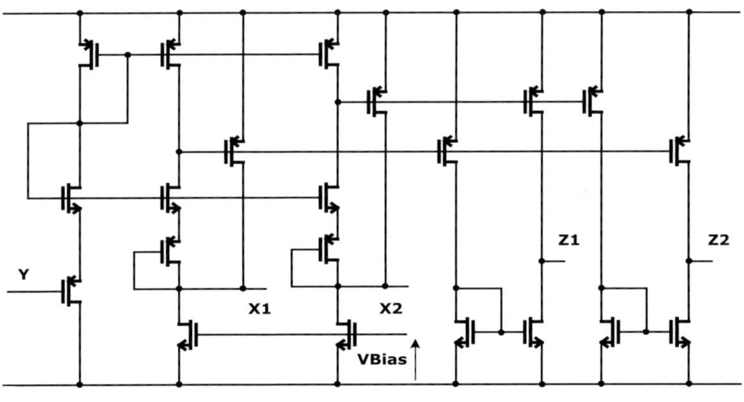

Figure 4.15– Class A DCCII implementation

In [10] the results for the circuit in figure 4.13 have been presented, and some applications, such as four quadrant multiplier and current-mode filter, are analysed too. Starting from the basic CCII block, a DCCII can be designed according to the topology reported in figure 4.16.
The Y nodes of two current conveyors have been connected together, so two low impedance X nodes have been obtained. These two X nodes are driven by the same Y voltage, as required by DCCII specifications (figure 4.14).

The block indicated as "current subtractor" is necessary to obtain the two Z node currents. Basically, it can be easily implemented by current mirrors. Its goal is to subtract reciprocally the currents flowing from Z nodes of the current conveyors, so I_{Z1} and I_{Z2} can be derived accordingly to DCCII characteristics.

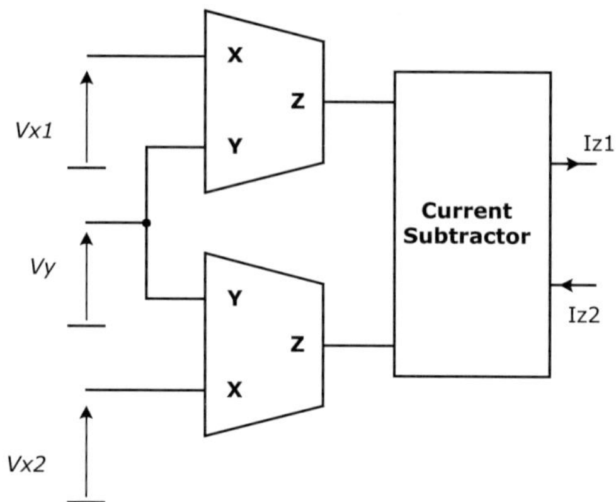

Figure 4.16 – CCII- based DCCII block. First solution

This solution may present some limitations due to the non-ideal performance of current mirrors forming the current subtractor. The circuit shown in the figure 4.17 allows to overcome this problem, because it performs the DCCII operation using only CCII blocks and three matched resistances (in the figure we will consider I_{z3} and I_{z4} currents as the output currents of the DCCII, which have been called I_{z1} and I_{z2} in the previous figures 4.14, 4.15 and 4.16).

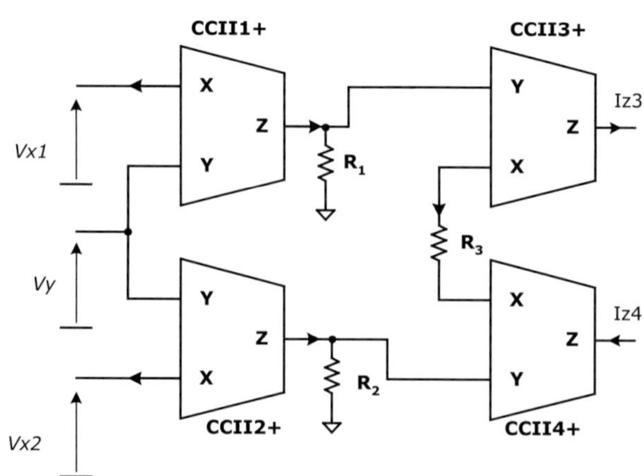

Figure 4.17 – CCII- based DCCII block. Second solution

Once more the current signals are converted, through R_1 and R_2 resistances, in two voltages, $R_1 I_{Z1}$ and $R_2 I_{Z2}$. These voltages, applied to CCII3 and CCII4, give the output currents I_{Z3} and I_{Z4}, as follows:

$$I_{Z3} = I_{X3} = \frac{V_{X3} - V_{X4}}{R_3} = \frac{V_{Y3} - V_{Y4}}{R_4} = \frac{V_{Z1} - V_{Z2}}{R_3} = \frac{R_1 I_{Z1} - R_2 I_{Z2}}{R_3} = \frac{R_1 I_{X1} - R_2 I_{X2}}{R_3} \tag{4.4}$$

$$I_{Z4} = I_{X4} = \frac{V_{X4} - V_{X3}}{R_3} = \frac{V_{Y4} - V_{Y3}}{R_3} = \frac{V_{Z2} - V_{Z1}}{R_3} = \frac{R_2 I_{Z2} - R_1 I_{Z1}}{R_3} = \frac{R_2 I_{X2} - R_1 I_{X1}}{R_3} \tag{4.5}$$

If all the resistances are equal and well matched,

$$I_{Z3} = I_{X1} - I_{X2} \; ; \; I_{Z4} = I_{X2} - I_{X1} \tag{4.6}$$

so I_{Z3} and I_{Z4} are the output currents of a DCCII, as stated in fig. 4.14. Implementing each CCII with a rail-to-rail LV LP current conveyor like that presented in figure 2.19 allows to obtain a rail-to-rail DCCII too. This gives an evident advantage with respect to the solution reported in figure 4.15.

4.2.2 Differential voltage CCII (DVCCII).

The differential voltage current conveyor (DVCCII) is characterised by two high-impedance input terminals (Y1 and Y2), one low-impedance node (X) and two high-impedance output nodes (Z1 and Z2). Its block scheme and matrix characteristics are summarised in figure 4.18 [9]:

$$\begin{bmatrix} V_X \\ I_{Y1} \\ I_{Y2} \\ I_{Z1} \\ I_{Z2} \end{bmatrix} = \begin{bmatrix} 0 & 1 & -1 & 0 & 0 \\ 0 & 0 & 0 & 0 & 0 \\ 0 & 0 & 0 & 0 & 0 \\ 1 & 0 & 0 & 0 & 0 \\ -1 & 0 & 0 & 0 & 0 \end{bmatrix} \begin{bmatrix} I_X \\ V_{Y1} \\ V_{Y2} \\ V_{Z1} \\ V_{Z2} \end{bmatrix}$$

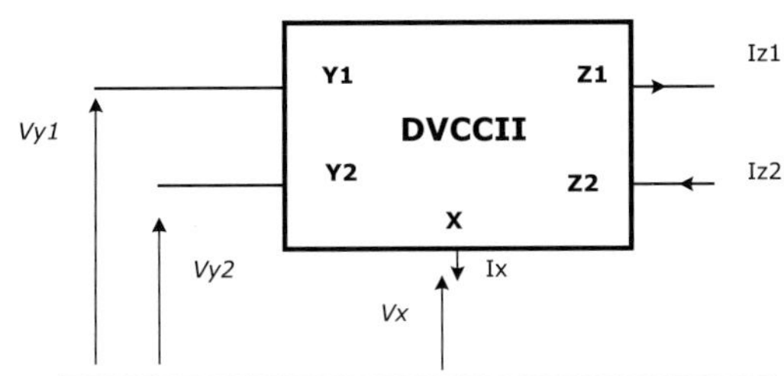

Figure 4.18 –DVCCII block representation and characteristics

In figure 4.19 a class A DCCII is shown [10], while figure 4.20 shows a DCCII implementation starting from CCII basic blocks. It has to be noted that the output currents I_{Z1} and I_{Z2} have been obtained by a dual output current conveyor.

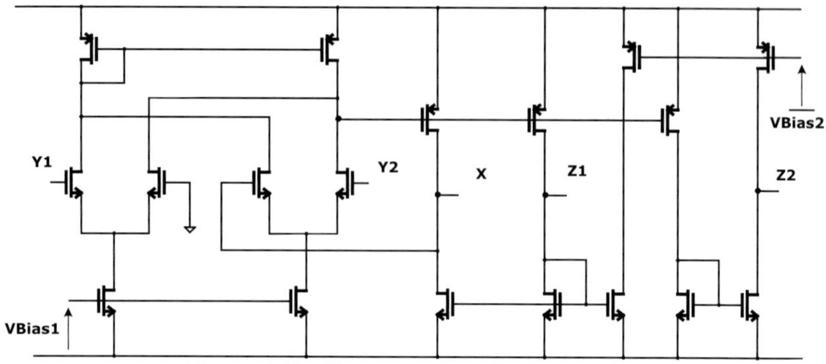

Figure 4.19 –Class A DVCCII implementation

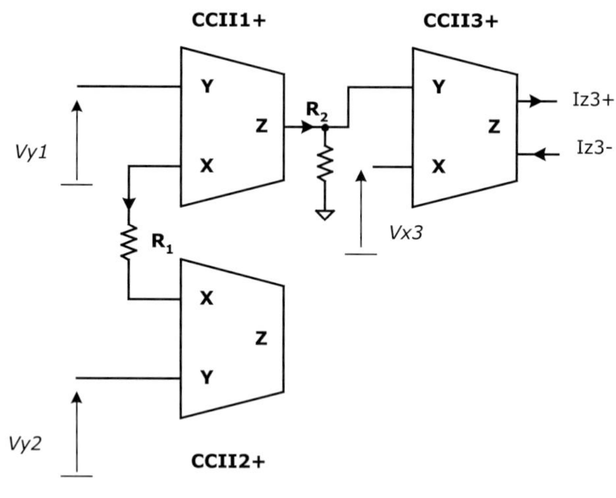

Figure 4.20 – CCII-based DVCCII block

The same principle applied to the DCCII can be used for the DVCCII, so designing a topology formed by basic CCIIs and resistances. The two voltages V_{Y1} and V_{Y2} force a current into R_1, which is mirrored to CCII1 Z node, thanks to the current conveyor characteristics. In fact we have:

$$V_X = V_{X3} = V_{Y3} = V_{Z1} = R_2\, I_{Z1} = R_2\, I_{X1} = \frac{R_2\,(V_{X1} - V_{X2})}{R_1} = \frac{R_2\,(V_{Y1} - V_{Y2})}{R_1} \qquad (4.7)$$

If $R_1 = R_2$,

$$V_X = V_{Y1} - V_{Y2}, \qquad (4.8)$$

according to the matrix characteristic in figure 4.18. The Z node currents have only to be equal to that flowing from X node, so a DOCCII has been used as the output current conveyor.

4.2.3 Fully differential CCII (FDCCII).

The natural evolution of DVCCII is a fully differential block (FDCCII), where each terminal has been doubled with respect to the original CCII. Its block scheme and matrix characteristic are summarised in figure 4.21 [11].

$$\begin{bmatrix} V_{X1} \\ V_{X2} \\ I_{Z1} \\ I_{Z2} \end{bmatrix} = \begin{bmatrix} 0 & 0 & -1 & 1 \\ 0 & 0 & 1 & -1 \\ 1 & 0 & 0 & 0 \\ 0 & 1 & 0 & 0 \end{bmatrix} \begin{bmatrix} I_{X1} \\ I_{X2} \\ V_{Y1} \\ V_{Y2} \end{bmatrix}$$

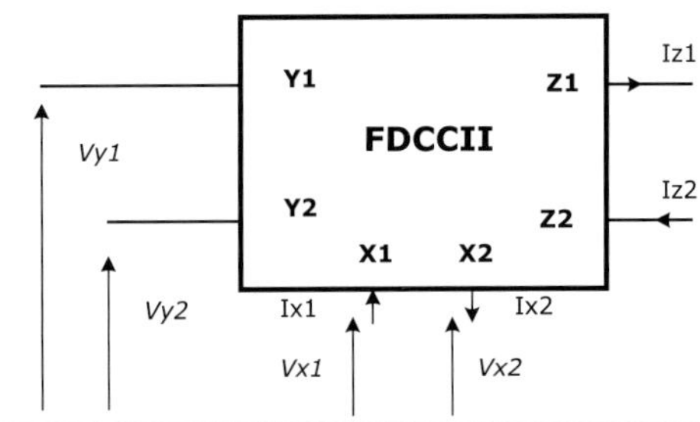

Figure 4.21 –FDCCII block representation and characteristics

The FDCCII may be considered as the most versatile building block that can be designed starting from the basic CCII. In fact, its topology can be thought as the "natural differential evolution" of the CCII idea. From figure 4.21 matrix description, it can be easily seen that each terminal of the CCII is replaced, in the FDCCII, by a couple of terminals, so obtaining a very useful block. In figure 4.22 a class AB implementation of FDCCII is shown [12].

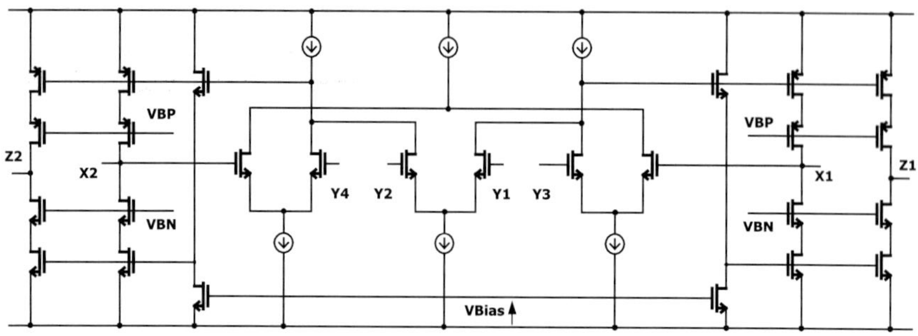

Figure 4.22 – Class AB FDCCII implementation

The solution in figure 4.22 presents two extra terminals with respect to the classical fully differential structure. This means that the topology is related to a more general matrix description, presented in (4.9).

$$\begin{bmatrix} V_{X1} \\ V_{X2} \\ I_{Z1} \\ I_{Z2} \end{bmatrix} = \begin{bmatrix} 0 & 0 & 1 & -1 & 1 & 0 \\ 0 & 0 & -1 & 1 & 0 & 1 \\ 1 & 0 & 0 & 0 & 0 & 0 \\ 0 & 1 & 0 & 0 & 0 & 0 \end{bmatrix} \begin{bmatrix} I_{X1} \\ I_{X2} \\ V_{Y1} \\ V_{Y2} \\ V_{Y3} \\ V_{Y4} \end{bmatrix} \tag{4.9}$$

Adding the current sensing principle to a DDA (Differential Difference Amplifier), it is possible to design a different FDCCII with very good performance and almost rail-to-rail output characteristics [11].

Also the FDCCII can be implemented through the use of basic CCIIs and resistances, as presented in figure 4.23. In this case too, any mismatch among R_1, R_2 and R_3 resistances will cause an error in the FDCCII behaviour. The drawbacks of this solution are represented by the high number of active components and the matching condition needed for the resistances employed but this is a very little price to pay, if compared to the fact that rail-to-rail differential current conveyors may be easily implemented. Moreover, each improvement in the CCII design in terms of power consumption, minimum supply voltage, bandwidth etc., automatically leads to differential circuits with better performance.

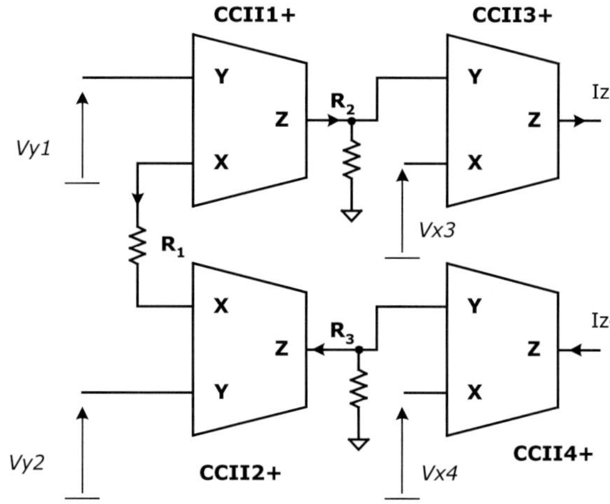

Figure 4.23 – CCII- based FDCCII block

The approach followed to derive a CCII-based DVCCII can be extended to the FDCCII shown in fig. 4.23. We have:

$$V_{X1} = V_{Y3} = R_2 I_{Z1} = R_2 I_{X1} = \frac{R_2 (V_{X1} - V_{X2})}{R_1} = \frac{R_2 (V_{Y1} - V_{Y2})}{R_1} \tag{4.10}$$

$$V_{X2} = V_{Y4} = R_3 I_{Z2} = R_3 I_{X2} = \frac{R_3 (V_{X2} - V_{X1})}{R_1} = \frac{R_3 (V_{Y2} - V_{Y1})}{R_1} \tag{4.11}$$

If $R_1 = R_2$,

$$V_{X1} = V_{Y1} - V_{Y2} \tag{4.12}$$

while if $R_1 = R_3$,

$$V_{X2} = V_{Y2} - V_{Y1}. \tag{4.13}$$

Once more all resistances have to be matched and parasitic components have to be taken into account for a more detailed analysis. Under these conditions eq.s (4.12) and (4.13) satisfy eq.(4.9).

4.2.4 Universal CCII (UCCII).

Starting from the first and second generation current conveyors, many types of new topologies have been designed during the past years. In [14], a universal current conveyor (UCCII) has been introduced with the aim to replace each current conveyor with its UCCII-based implementation.
In fact, it could be demonstrated that each block described in the previous paragraphs, from DOCCII to DVCCII, can be obtained from the universal current conveyor itself, whose characteristics, in matrix form and block, are pictured in figure 4.24.
The implementation of such a block is based on that presented in figure 4.19 for the DVCCII. Some extra Z nodes have been introduced. In order to obtain all the possible CCII evolutions, each Z node presents its negative one. For example, considering Y_1 as the Y node and Z_1 or Z_2 as the Z node, a classic CCII+ is easily derived. All the other blocks can be implemented and, moreover, some novel topologies have been also introduced.

$$\begin{bmatrix} I_{Y1} \\ I_{Y2} \\ I_{Y3} \\ V_X \\ I_{Z1} \\ I_{Z2} \\ I_{\overline{Z1}} \\ I_{\overline{Z2}} \end{bmatrix} = \begin{bmatrix} 0 & 0 & 0 & 0 & 0 & 0 & 0 & 0 \\ 0 & 0 & 0 & 0 & 0 & 0 & 0 & 0 \\ 0 & 0 & 0 & 0 & 0 & 0 & 0 & 0 \\ 1 & -1 & 1 & 0 & 0 & 0 & 0 & 0 \\ 0 & 0 & 0 & 1 & 0 & 0 & 0 & 0 \\ 0 & 0 & 0 & 1 & 0 & 0 & 0 & 0 \\ 0 & 0 & 0 & -1 & 0 & 0 & 0 & 0 \\ 0 & 0 & 0 & -1 & 0 & 0 & 0 & 0 \end{bmatrix} \begin{bmatrix} V_{Y1} \\ V_{Y2} \\ V_{Y3} \\ I_X \\ V_{Z1} \\ V_{Z2} \\ V_{\overline{Z1}} \\ V_{\overline{Z2}} \end{bmatrix}$$

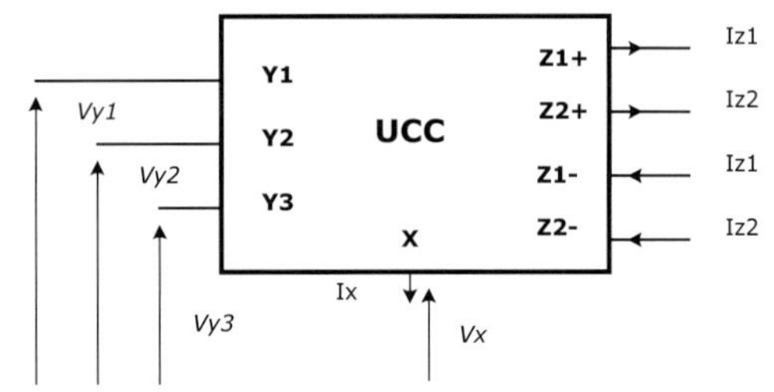

Figure 4.24 –UCCII block representation and characteristics

As seen for the previously introduced blocks, the UCCII can be implemented using basic CCII blocks, as shown in figure 4.25. We have :

$$V_X = V_{X4} = V_{Y4} = R_2\,(I_{Z1} + I_{Z3}) = \frac{R_2\,(V_{X1} - V_{X2})}{R_1} + \frac{R_2\,V_{X3}}{R_3} = \frac{R_2\,(V_{Y1} - V_{Y2})}{R_1} + \frac{R_2\,V_{Y3}}{R_3} \tag{4.14}$$

If all the resistances are equal,

$$V_X = V_{Y1} - V_{Y2} + V_{Y3} \tag{4.15}$$

according to the matrix characteristic.

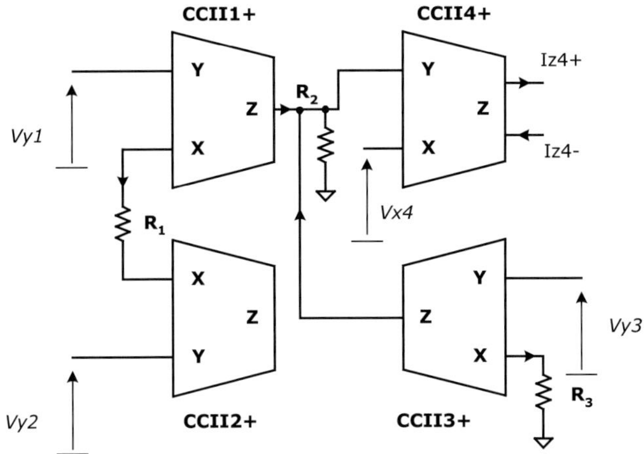

Figure 4.25 – CCII- based UCCII block

In conclusion, in this Chapter we have presented all the evolutions of the basic second generation current conveyor towards its differential versions, which consist of multiple input and output terminals. These circuits, which can be obtained both from modification of the basic CCII block at transistor level and through opportune external connections of more CCIIs and passive components, can be utilized in a number of interesting applications, some of which will be shown in the next chapter.

References.

[1] G.Di Cataldo, G.Ferri, S.Pennisi. Active capacitance multipliers using current conveyors, Proceedings of International Symposium of Circuits and Systems, 1998; Monterey, U.S.A.

[2] G.Ferri, S.Pennisi, A 1.5 V Current-Mode Capacitance Multiplier. Proceedings of International Conference on Microelectronics, 1998; Monastir, Tunisia.

[3] G.Ferri, S.Pennisi, S.Sperandii, A low voltage CMOS 1-Hz low pass filter. Proceedings of International Conference on Electronics, Circuits and Systems, 1999; Cyprus.

[4] P.De Laurentiis, G.Ferri, G.Palumbo, S.Pennisi, A low-pass 1-Hz 2V-supply current-conveyor based filter, Proceedings of European Circuits and Systems Conference, 1999; Bratislava, Slovakia.

[5] G.Ferri, N.Guerrini. High valued passive element simulation using low-voltage low-power current conveyors for fully integrated applications. IEEE Transactions on Circuits and Systems II. nr.4; vol.48; 2001; pp.405-409.

[6] H. Barthelemy, A. Fabre. A second generation current controlled conveyor with negative intrinsic resistance. IEEE Transactions on Circuit and Systems I. vol. 49; 2002; pp. 63-65.

[7] A. Fabre, O. Saaid, F. Wiest, C. Boucheron. High-frequency high-Q BiCMOS current-mode bandpass filter and mobile communication application. IEEE Journal of Solid-State Circuits. nr. 4; vol. 33; 1998; pp. 614-624.

[8] A. Fabre. Third generation current conveyor: a new helpful active element. Electronics Letters. nr. 5; vol. 31; 1995; pp. 338-339.

[9] H. O. Elwan, A. M. Soliman. Novel CMOS differential voltage current conveyor and its applications. IEE Proceedings - Circuits Devices and Systems. nr. 3; vol. 144; 1997; pp. 195-200.

[10] H. O. Elwan, A. M. Soliman. CMOS differential current conveyors and applications for analog VLSI. Analog Integrated Circuits and Signal Processing. nr. 11; 1996; pp. 35-45.

[11] H. A. Alzaher, H. O. Elwan, M. Ismail. CMOS fully differential second-generation current conveyor. Electronics Letters. nr. 13; vol. 36; 2000; pp. 1095-1096.

[12] A.A.El-Adawy, A.M. Soliman, H.O.Elwan. A novel fully differential current conveyor and applications for analog VLSI. IEEE Transactions on Circuit and Systems-II. nr. 4; vol. 47; 2000; pp. 306-313.

[13] D. Becvar, K. Vrba, V. Zeman, V. Musil. Novel universal active block: a universal current conveyor. Proceedings of the IEEE International Symposium on Circuits and Systems, 2000; Geneva, Switzerland.

CHAPTER V

LV LP CCII APPLICATIONS

Second generation current conveyors can be employed in a very large number of different applications. Owing to its flexibility and versatility, CCII can be seen as a real competitor for the operational amplifier. In fact, in several applications, current conveyors allow to obtain better results than op-amp ones, but CCII can replace op-amp only when it is proved that *all* the signal processing functions, usually implemented with op-amps, can be performed with similar or better characteristics through the use of CCIIs. That's why in the following sections a number of simple applications with CCIIs will be investigated and, in some cases, compared to the related op-amp-based ones. In this sense, there are two different approaches. The first approach utilises them as substituting blocks of the op-amps, while the second one proposes a different design philosophy, named *current-mode*, and introduces CCII itself as the basic building block in this new environment.

5.1 BASIC APPLICATIONS

In the first chapter, where the general feedback theory has been presented, it has been shown that the best basic block which implements a voltage amplifier is the VCVS (Voltage Controlled Voltage Source) [1,2]. The analysis of the basic applications will start with the comparison between op-amp-based and CCII-based voltage amplifiers.
In figure 5.1 a traditional op-amp-based voltage amplifier is pictured. Assuming the op-amp as ideal - in particular, considering infinite gain at each operating frequency and infinite input impedance – the relation between input and output voltages is expressed by the following equation:

$$V_{OUT} = A*V_{IN} = \left(1 + \frac{R_2}{R_1}\right)*V_{IN} \tag{5.1}$$

Considering a finite gain G for the op-amp, eq.(5.1) becomes :

$$V_{OUT} = A*V_{IN} = \frac{\left(1+\frac{R_2}{R_1}\right)}{1+\frac{\left(1+\frac{R_2}{R_1}\right)}{G}}*V_{IN} \tag{5.2}$$

Usually, G has a single-pole behaviour, so, when the frequency is raised, the effective gain of the operational amplifier decreases, so leading to a greater difference between theoretical and ideal behaviour.

The op-amp-based traditional voltage amplifier presents the heavy limitation of the non ideal gain of the basic block, especially at higher frequencies. This limitation can be expressed in terms of the error introduced by non-ideal op-amps, which is defined as follows :

$$Error\% = \frac{\left(1+\frac{R_2}{R_1}\right)}{G+\left(1+\frac{R_2}{R_1}\right)}*100 \tag{5.3}$$

Considering eq.(5.3), it is clear that if the low frequency op-amp gain is equal to 60 dB, the voltage gain error will be 0.1% ; but this error raises up to a value of 90.9% at the GBW frequency, where G=1, considering a typical ratio R_2/R_1 of 10.

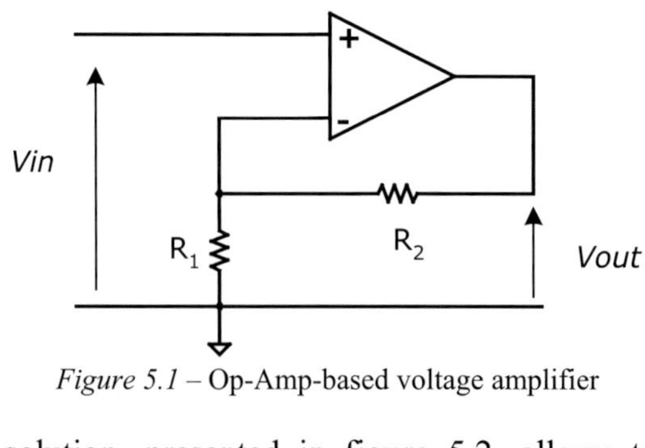

Figure 5.1 – Op-Amp-based voltage amplifier

The CCII-based solution, presented in figure 5.2, allows to overcome these drawbacks. In fact, the circuit behaviour is very close to the expected one if the voltage at X node is equal to that applied to Y node.

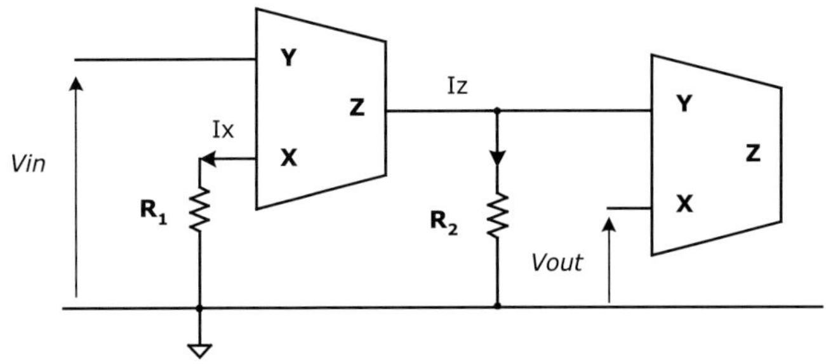

Figure 5.2 – CCII-based voltage amplifier

As shown in the previous chapters, it is possible to have good performance implementing a good CCII input voltage buffer, through the use of a differential pair. In this case, α parameter is close to its ideal unitary value for a frequency range much larger with respect to that where op-amps present a very high gain. This feature has the result to allow a wider operating frequency range for all CCII-based circuits. The latter consideration is usually expressed by the fact that CCIIs do not show the limitation of a constant GBW product.

The voltage amplifier shown in figure 5.2 is implemented through the use of two CCII blocks. The second CCII performs the buffering operation, so it can be avoided either when R_2 resistance can be considered the load of the voltage amplifier or when it is much lower than the load resistance. In this last case, in fact, the load is not affecting the amplifier voltage gain.

The use of two CCIIs does not increase the complexity of the overall circuit, in terms of number of transistors (an op-amp requires internally much more transistors than a CCII) and, consequently, power dissipation can be kept at reduced values.

The voltage gain expression for the circuit proposed in figure 5.2 comes from the following relationship:

$$V_{OUT} = V_{X2} = V_{Y2} = R_2 I_{Z1} = R_2 I_{X1} = R_2 \frac{V_{X1}}{R_1} = R_2 \frac{V_{IN}}{R_1} = \frac{R_2}{R_1} V_{IN} \tag{5.4}$$

This expression is valid assuming both CCIIs as ideal, and the input CCII as a positive current conveyor. Using a negative CCII, an inverting voltage amplifier is obtained. This fact is particular important because it is impossible to perform both functions with op-amps, which are able to implement, typically, only inverting topologies.

If CCII non-idealities (α and β) are taken into account, we can write :

$$V_{OUT} = V_{X2} = \alpha V_{Y2} = \alpha_2 R_2 I_{Z1} = \alpha_2 \beta_1 R_2 I_{X1} = \alpha_2 \beta_1 R_2 \frac{V_{X1}}{R_1} = \alpha_1 \alpha_2 \beta_1 \frac{R_2}{R_1} V_{IN} \tag{5.5}$$

α and β parameters are usually very close to unit, so the error results to be quite low.

Let us now consider other different applications of operational amplifiers, so to be compared to the same functions implemented through CCIIs. In figure 5.3 an op-amp-based current amplifier is shown.

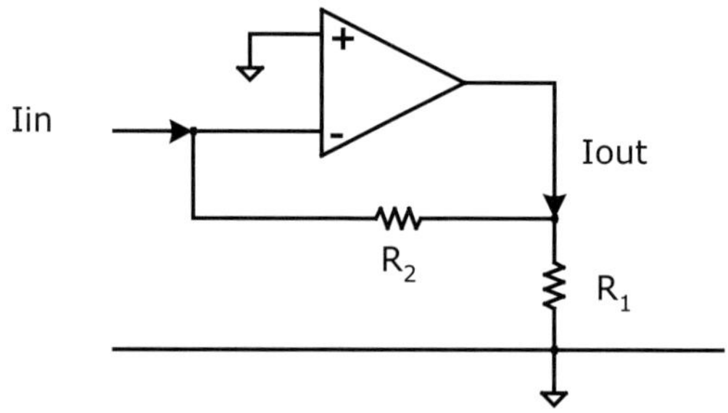

Figure 5.3 – Op-Amp-based current amplifier

The current gain expression is very simple when the op-amp is ideal. In fact, in this case we can write :

$$I_{OUT} = -\left(1 + \frac{R_2}{R_1}\right) I_{IN} \tag{5.6}$$

but, if a finite gain G is considered, we have :

$$I_{OUT} = -\left(\frac{G(R_1 + R_2) - R_1}{GR_1 - R_1}\right) I_{IN} = -\left(\frac{(G-1)R_1 + GR_2}{(G-1)R_1}\right) I_{IN} \tag{5.7}$$

If we consider an amplifier with a high gain it is possible to recover eq.(5.6). Eq.(5.7) confirms that a voltage amplifier with a current gain feedback should not present the gain bandwidth limitation, because the current gain does not depend on the op-amp gain. Anyway, we must consider the fact that increasing the frequency, the op-amp voltage gain tends to 1, so causing a strong increment in the current gain (for G=1, eq.(5.7) tends ideally to infinite).

However, the main problem in this solution is related to the impedance levels. From feedback theory, summarised in table 1.1, it is clear that the feedback applied to a current amplifier reduces the input impedance and increases the output one by a factor exactly equal to the loop gain. So, if a 1 MΩ output resistance is required, starting from the few Ωs of the output impedance of a common op-amp, a 100-120 dB loop gain is needed. This means that even if there is not the frequency limitation due to constant GBW product, it is quite hard to meet the impedance constraints.
Moreover, R_2, which is used to obtain the desired current gain, is the load resistance. Then, a different implementation of a current amplifier can be performed through the use of a CCII (figure 5.4).

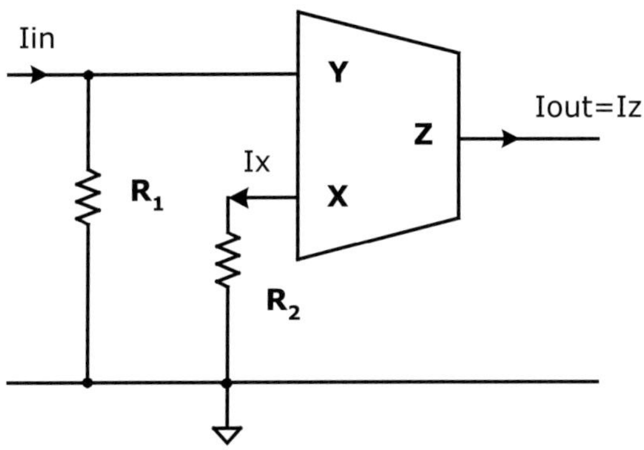

Figure 5.4 – CCII-based current amplifier

In this case, only one current conveyor is needed because no output voltage buffer operation is required. Firstly, the input current is converted into an input voltage, applied to Y node. The X node voltage obtained in this way forces, into R_2, a current which flows from high impedance Z node.

$$I_{OUT} = I_Z = I_X = \frac{V_X}{R_2} = \frac{V_Y}{R_2} = R_1 \frac{I_{IN}}{R_2} = \frac{R_1}{R_2} I_{IN} \tag{5.8}$$

Also in this case, CCII non-idealities, in terms of α and β parameters, introduce an error which can be taken into account as follows:

$$I_{OUT} = I_Z = \beta I_X = \beta \frac{V_X}{R_2} = \alpha\beta \frac{V_Y}{R_2} = \alpha\beta \frac{R_1}{R_2} I_{IN} \tag{5.9}$$

The impedance problems seen for the previous op-amp-based solution are overcome because the input impedance is practically equal to R_1, while the output one is the CCII output impedance, that is high and not related to the imposed current gain. In addition to that, a current-gained CCII (CGCCII, see chapter 4) may be also used as a building block, giving a possible additive current gain.
Let us consider now the transconductance amplifier. The op-amp based one, reported in figure 5.5, shows a gain equal to the inverse of the load resistance R_1. In this case, the error due to non ideal op-amp behaviour is not the most important problem, because there is again the problem of the output impedance level.

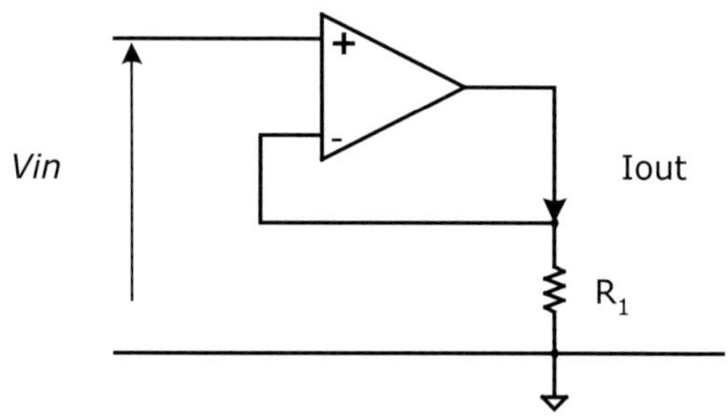

Figure 5.5 – Op-Amp-based transconductance amplifier

On the contrary, CCII-based solution, pictured in figure 5.6, shows an easily tunable gain (equal to $1/R_1$) and good impedance levels at both input (Y) and output (Z) nodes.

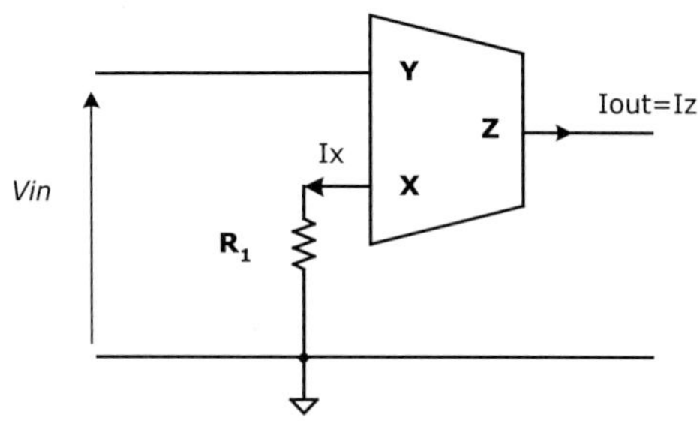

Figure 5.6 – CCII-based transconductance amplifier

Also in this case the circuit can be more flexible if a CGCCII is used instead of a simple CCII. In fact, it is possible to limit the power consumption reducing I_X current (through an increase of R_1 value), but maintaining the same gain.

In this manner, the CGCCII-based transconductance amplifier has a gain of $1/(KR_1)$, where K is the ratio between I_Z and I_X currents and R_1 is the resistance connected to X node.
The last basic amplifier configuration is the transimpedance amplifier, whose traditional implementation with op-amp is presented in figure 5.7.

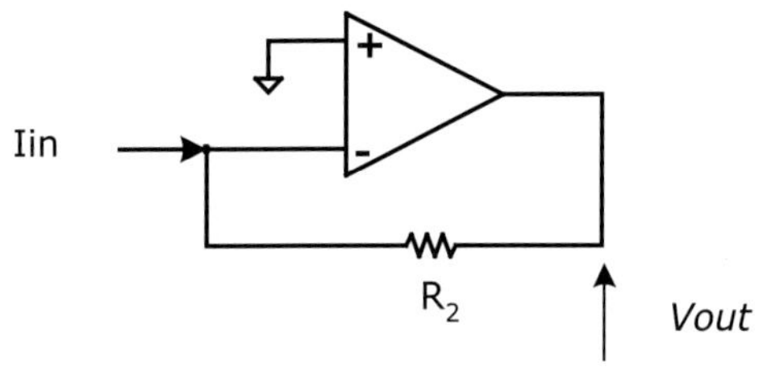

Figure 5.7 – Op-Amp-based transimpedance amplifier

Also for what concerns this topology, there is not the problem linked to the constant GBW product (see table 1.1), but the limitation is again due to the impedance levels. In fact, the input impedance should be as low as much possible, but op-amps have very large input impedances. The feedback effect helps to reduce this value by a factor related to loop gain, but a strong reduction can be obtained only at low frequencies, where the op-amps show very high voltage gains.
The equivalent CCII-based solution, shown in figure 5.8, allows to choose the desired gain without affecting the impedance levels. In fact, the low impedance levels are ensured by CCII characteristics, and are not related to feedback effects.

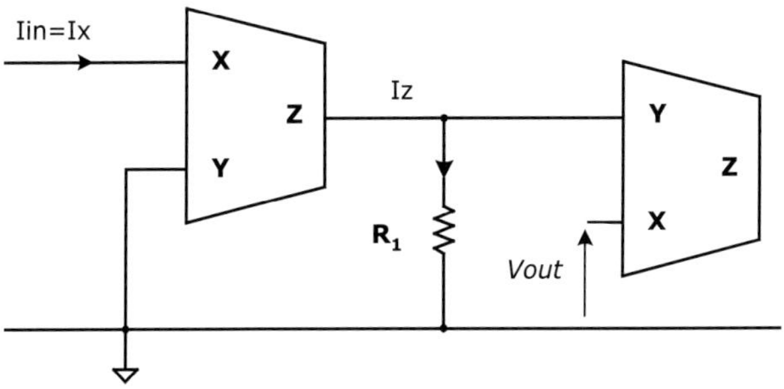

Figure 5.8 – CCII-based transimpedance amplifier

From the previously presented results, it appears clear that current conveyors are very powerful building blocks, allowing analog designers to implement the four basic amplifier configurations with performance that are similar or, in some cases, improved with respect to their opamp-based counterparts. In the next, other CCII-based applications will be considered.
In figure 5.9 a CCII-based current adder is reported. (It is easy to extend the topology to the case of a higher number of input currents.)

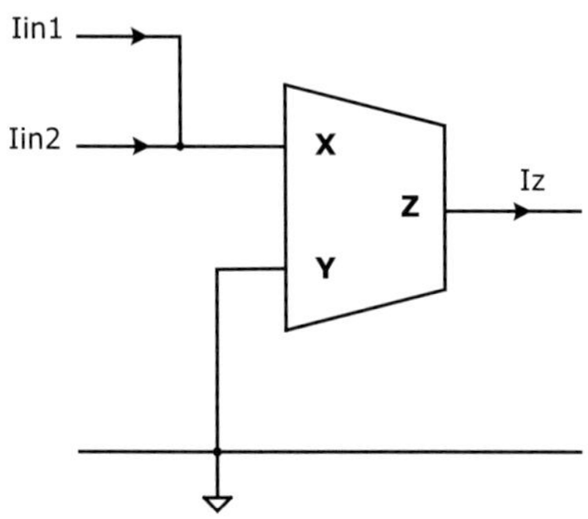

Figure 5.9 – CCII-based current adder

Thanks to the fact that, in the CCII+, $I_Z = I_X$, a current adder is implemented simply connecting more than a current source to the low impedance input node X. The only source of error is linked to β parameter which is not exactly equal to its ideal unitary value.
This solution can be modified to design a CCII-based voltage adder, as depicted in figure 5.10.

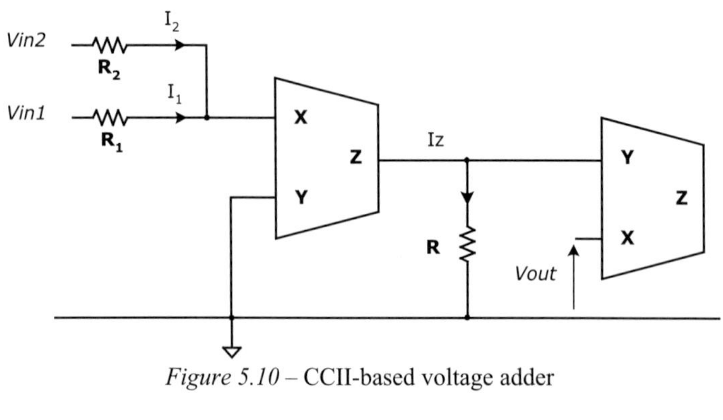

Figure 5.10 – CCII-based voltage adder

Input voltages are converted into currents through R_1 and R_2 resistors. The first CCII performs the current adder action seen before. Then, the output current is converted again into a voltage through the use of R resistor. In order to obtain a low output impedance level, a second CCII has been used as a voltage buffer.
In figure 5.11, a simple current integrator is presented.

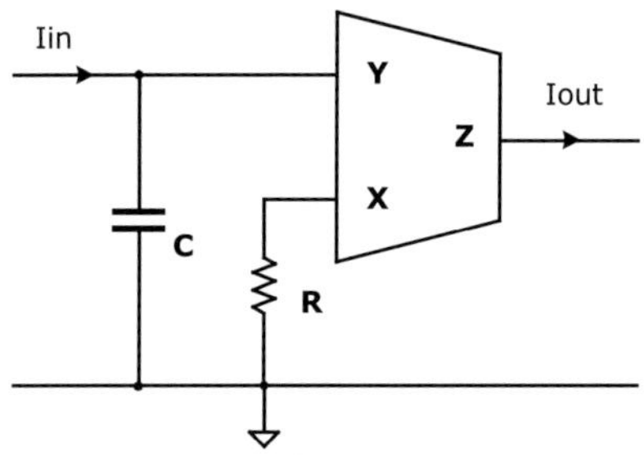

Figure 5.11 – CCII-based current integrator

The expression of the output current is easily evaluated considering the ideal CCII characteristics, as follows:

$$I_{OUT} = I_Z = I_X = \frac{V_X}{R} = \frac{V_Y}{R} = \frac{\frac{1}{sC} I_{IN}}{R} = \frac{I_{IN}}{sCR} \tag{5.10}$$

Also in this case it is possible to take into account, in eq.(5.10), all the CCII non idealities in a very simple way.
Swapping the two passive elements, a current differentiator is obtained, as pictured in figure 5.12.

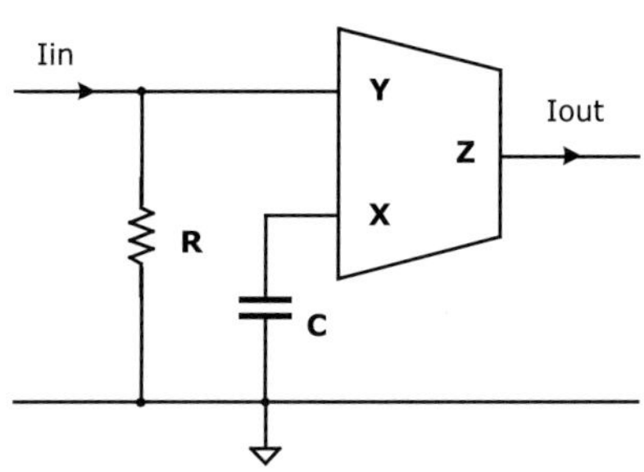

Figure 5.12 – CCII-based current differentiator

In this case we can write:

$$I_{OUT} = I_Z = I_X = \frac{V_X}{\frac{1}{sC}} = sCV_X = sCV_Y = sCRI_{IN} \quad (5.11)$$

CCII is not only a block suitable for current-based applications, but allows to implement also voltage-based applications.

In figure 5.13 a CCII-based voltage integrator is proposed. The second current conveyor ensures again the voltage buffer operation at the output.

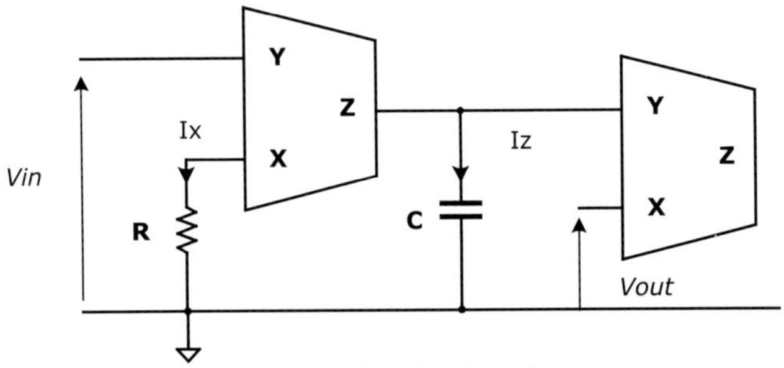

Figure 5.13 – CCII-based voltage integrator

A straightforward analysis for the circuit of figure 5.13 brings to the following expression for the output voltage:

$$V_{OUT} = V_{X2} = V_{Y2} = \frac{I_{Z1}}{sC} = \frac{I_{X1}}{sC} = \frac{V_{X1}}{sCR} = \frac{V_{Y1}}{sCR} = \frac{1}{sCR} V_{IN} \quad (5.12)$$

Also in this case the passive elements can be swept to design a voltage differentiator, shown in figure 5.14.

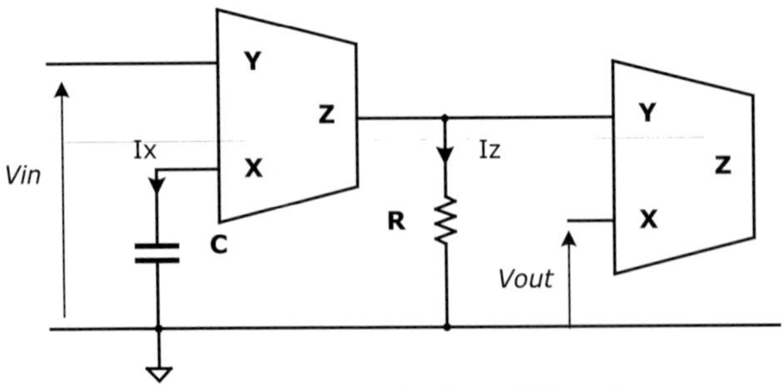

Figure 5.14 – CCII-based voltage differentiator

The output voltage expression (eq.(5.13)) confirms that the circuit behaves as required. In fact:

$$V_{OUT} = V_{X2} = V_{Y2} = RI_{Z1} = RI_{X1} = \frac{V_{X1}}{\frac{1}{sC}}R = sRCV_{X1} = sRCV_{Y1} = sRCV_{IN} \tag{5.13}$$

With respect to the op-amp counterpart, the proposed CCII-based integrator and differentiator are non-inverting configurations. Traditional inverting configurations can be implemented simply changing the type of CCII !
In the next paragraphs other CCII-based applications, together with their characteristics, will be presented.

5.2 IMPEDANCE SIMULATORS.

Thanks to their particular characteristics, CCIIs can be widely used in the implementation of circuits able to convert impedances. Negative impedances converters in general, and FDNR (Frequency Dependent Negative Resistance) in particular, are usually implemented by CCIIs instead of op-amps because of their better performance.

5.2.1 Impedance Converters.

In figure 5.15 a simple CCII-based negative impedance converter has been reported [2].

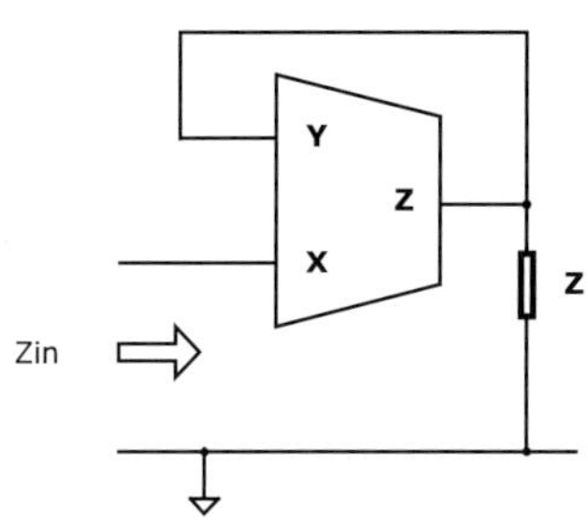

Figure 5.15 – Current controlled negative impedance converter

Using a positive CCII, the current injected at X node will flow into Z impedance, so imposing a negative voltage at Y node. This value is also present at X node, where the equivalent impedance is equal to –Z, as shown in eq.(5.14).

$$Z_{IN} = -\frac{V_X}{I_X} = -\frac{V_Y}{I_Z} = -\frac{V_Z}{I_Z} = -Z \tag{5.14}$$

If we consider, as a source, a voltage generator instead of a current generator, the solution presented in figure 5.16 can be employed, where the equivalent impedance at Y node is now considered.

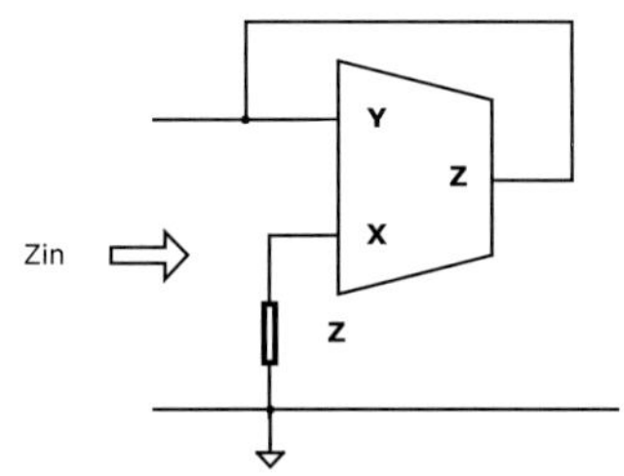

Figure 5.16 – Voltage controlled negative impedance converter

The operating principle is exactly the same and the resulting input impedance at input node Y is still equal to –Z.

$$Z_{IN} = -\frac{V_Y}{I_Z} = -\frac{V_Y}{I_X} = -\frac{V_X}{I_X} = -Z \tag{5.15}$$

The two solutions proposed in figure 5.15 and 5.16 allow the simulation of a grounded negative impedance. In order to implement a floating negative impedance, it is necessary to introduce a second CCII. A possible implementation is reported in figure 5.17. We notice that this topology has been obtained simply doubling that shown in figure 5.16, so the resulting input impedance is still equal to –Z.
Using CCIIs, it is also possible to design circuits having general purpose features, such as the GIC (general impedance converter), whose block scheme is presented in figure 5.18.

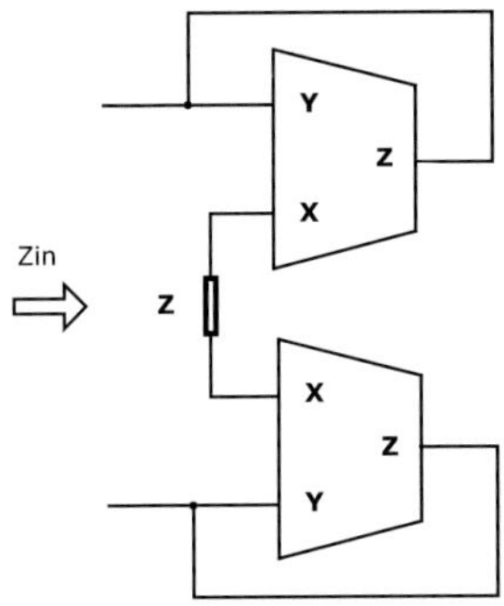

Figure 5.17 – Floating negative impedance converter

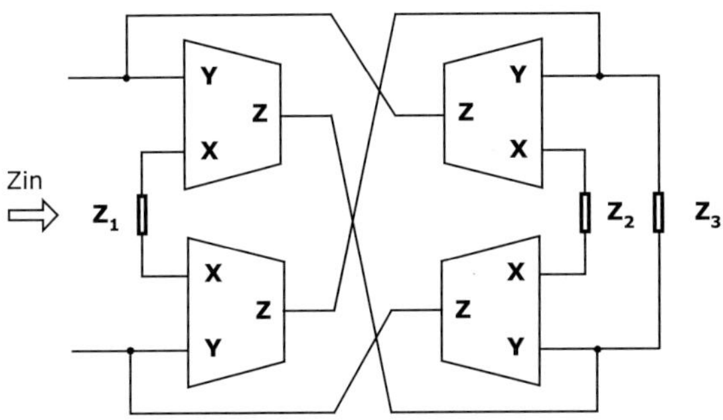

Figure 5.18 – General Impedance Converter (GIC)

The equivalent impedance seen at the input of the circuit reported in figure 5.18 is given by:

$$Z_{IN} = \frac{Z_1 Z_2}{Z_3} \tag{5.16}$$

Changing the value and/or the kind of impedances inserted in the GIC, it is possible to obtain different equivalent input impedances. A negative impedance converter is built-up simply replacing the two current conveyors whose X nodes are connected to Z_2 by two negative current conveyors. This means that both positive and negative CCIIs have to be used. In integrated applications it could be a great advantage to implement functions using basic blocks of the same type. To this aim, it is possible to modify the solution presented in figure 5.18 in order to design a negative impedance converter based only on CCII-. The block scheme is shown in figure 5.19.

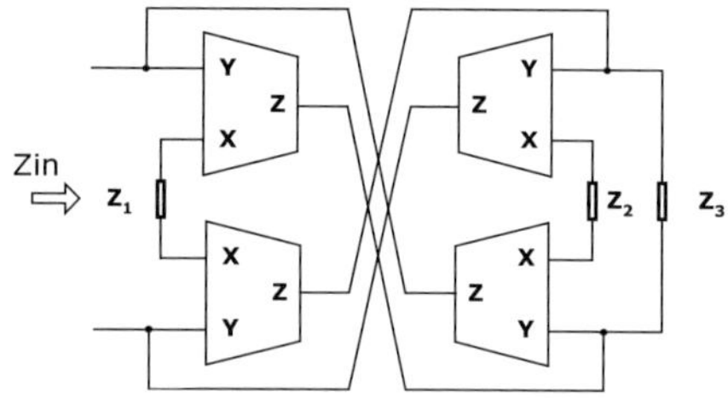

Figure 5.19 – Negative CCII-based General Impedance Converter

5.2.2 Capacitance Multiplication

Using LV LP CCIIs, it is possible to simulate the frequency behaviour of high capacitances and inductances by utilising only elements suitable for an integrated implementation.

Large capacitances need large silicon areas in integrated applications, so their implementation is limited to small values. If large capacitance are needed, there are only two solutions, using either external capacitances or designing circuits able to simulate the behaviour of such passive components [3].

In particular, it is possible to implement high capacitance values from smaller ones by performing a capacitance multiplication. This can be obtained using the CGCCII which shows a current gain between Z and X terminal, as in the circuit of figure 5.20 [4, 5, 6].

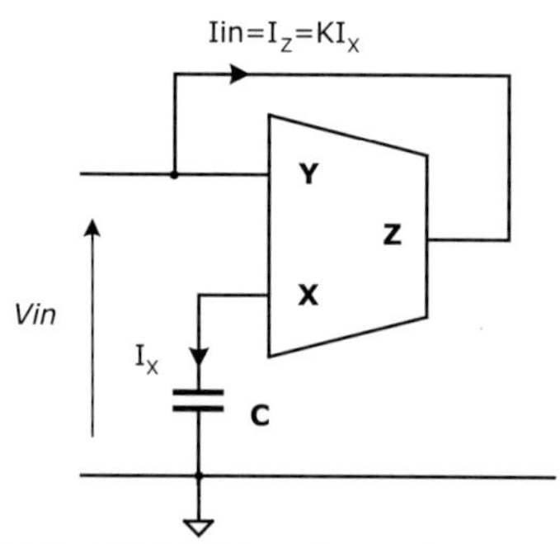

Figure 5.20 – CGCCII-based capacitance multiplier

If the current flowing from Z node is K times greater than that flowing from X node, the impedance at X node is K times higher than that seen from Y node, as clarified by the following formulas:

$$Z_{IN} = \frac{V_{IN}}{I_{IN}} = \frac{V_Y}{I_{IN}} = \frac{V_Y}{I_Z} = \frac{V_Y}{KI_X} = \frac{V_Y}{K\frac{V_X}{Z_C}} = \frac{V_Y}{K\frac{V_Y}{Z_C}} = \frac{Z_C}{K}$$

(5.17)

from which

$$Z_{IN} = \frac{1}{sCK}$$

(5.18)

being $Z_c = 1/sC$.

In this manner, we have, at Y node, an equivalent impedance whose value is K times smaller than X_C, which means a capacitance K times higher than C. Hence, the capacitance multiplication is performed.

If a real CCII is used as a capacitance multiplier, implementing the scheme in the figure 5.20, the equivalent impedance at Y node Z_{IN} is given by:

$$Z_{IN} = \frac{R_Z(1 + sCR_X)}{1 + s(C_Z R_Z + CR_X + KCR_Z + C_Y R_Z) + s^2 CR_X R_Z (C_Z + C_Y)}$$

(5.19)

From eq.(5.19), if $R_X = C_Z = C_Y = 0$ and $R_Z = \infty$ (ideal case), we recover eq.(5.18).

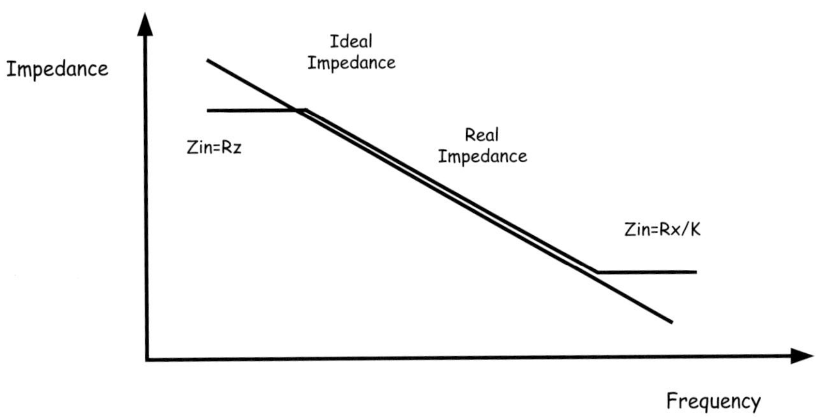

Figure 5.21 – Ideal and real (or non ideal) impedance behaviour (for $C_Z = C_Y = 0$) at CGCCII based capacitance multiplier input

Figure 5.21 shows the ideal and non ideal impedance behaviour (neglecting the high frequency pole, it means considering $C_Z=C_Y=0$) at CGCCII based capacitance multiplier input. Simulation results suggest that an acceptable agreement between theoretical and obtained behaviour is not possible for more than three or four frequency decades. From eq.(5.19), it can be found, neglecting the second order term, that for low frequencies the equivalent impedance is equal to R_Z, while at high frequencies it tends to R_X/K (always considering $C_Z=C_Y=0$), so the frequency range is limited by these two values and can be adjusted modifying R_X and R_Z values, through an opportune modification of CCII output stages biasing currents.
Since the capacitive impedance decreases when the frequency is growing, if an operating frequency range towards higher frequencies is needed, then R_X and R_Z have to be reduced. Vice versa if a range towards lower frequencies is required, R_X and R_Z have to be increased. The relative percentage error in capacitance multiplication has resulted to be lower than 3% for each C and K condition.

5.2.3 Inductance Simulation

In the previous section, the capacitance multiplication has been introduced in order to face the problem represented by the large silicon area required by high capacitances.
The implementation of inductances offers a similar problem to the designers. It is not possible to implement an inductance on a chip starting from a standard CMOS fabrication process, so circuits able to emulate an inductive behaviour have been proposed during the years.
The most famous inductance simulator (shown in figure 5.22) was proposed by Antoniou and utilizes two op-amps and five passive elements to obtain an equivalent impedance given by [7]:

$$Z_{IN} = \frac{V_{IN}}{I_{IN}} = \frac{sCR_1R_2R_4}{R_3} \tag{5.20}$$

This solution has particular features, such as the fact that it is not very sensitive to non ideal op-amp behaviour. Moreover, large inductances can be easily obtained with low values of resistances, but the topology presents the heavy drawback to implement only grounded equivalent inductances, while typically these are series components in passive networks. In figure 5.23 an alternative solution for the simulation of a grounded inductance based on CCIIs is presented [6].

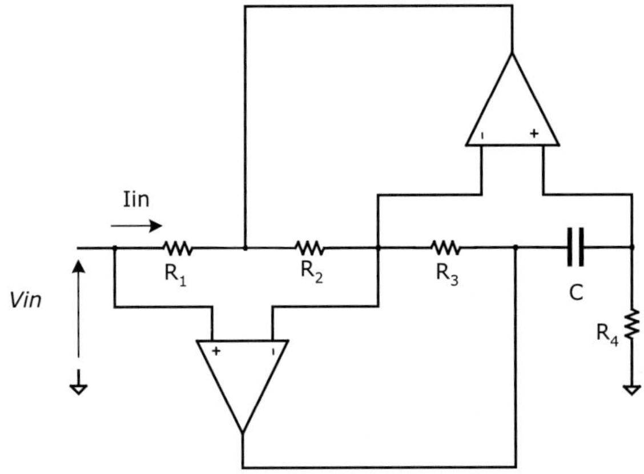

Figure 5.22 – Antoniou's inductance simulator

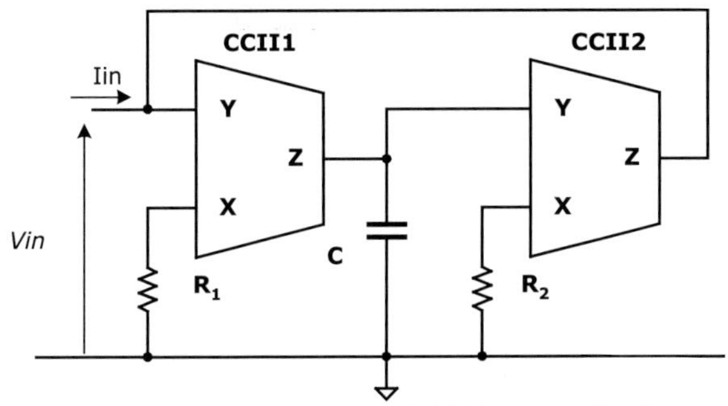

Figure 5.23 – CCII-based grounded inductance simulator

Considering ideal CCIIs, the equivalent inductance value is R_1R_2C, but, if the non-ideal active components are taken into account, a slightly different value in a limited frequency operating range is obtained (see figure 5.24), according to the following formula :

$$Z_{IN} = \frac{(R_1 + R_X)(R_2 + R_X)(1 + sR_Z(C_Z + C_Y + C))}{R_Z} \tag{5.21}$$

Different values of simulated grounded inductances can be obtained using different values for R_1,R_2 and C. The percentage error is lower than 3%. From eq.(5.21) it follows that the series resistance and the Q factor are respectively:

$$R_S = \frac{(R_1 + R_X)(R_2 + R_X)}{R_Z} \tag{5.22}$$

$$Q = \omega \frac{L}{R_S} \cong \omega R_Z (C_Z + C_Y + C) \tag{5.23}$$

As well known, the best inductances show a low series resistance and a high quality factor. From eq.s (5.22) and (5.23), it comes that R_1 and R_2 have to be low, while C has to be high. Unfortunately, the choice of a high capacitance leads to have operating ranges towards the low frequencies, so a trade-off has to be made during the design process.

With respect to op-amp based solution, a smaller number of passive components is required, while larger inductances need larger values for passive elements. If a circuit completely suitable for integration is needed, the capacitor C, shown in figure 5.23, has to be replaced by a capacitance multiplier like the one presented in the previous section. In this case, a class A CCII-based capacitance multiplier has been used, because it allows to reach a better precision, in this operation, with respect to the same solution implemented with a class AB CCII. The overall error is obviously increased but is still lower than 7-8%. Employing an high value capacitance through a capacitance multiplier leads to obtain a worst precision in inductance simulation but the circuit is now suitable for a fully integrated implementation. It must be noted that the operating frequency range is not affected by the use of a capacitance multiplier.

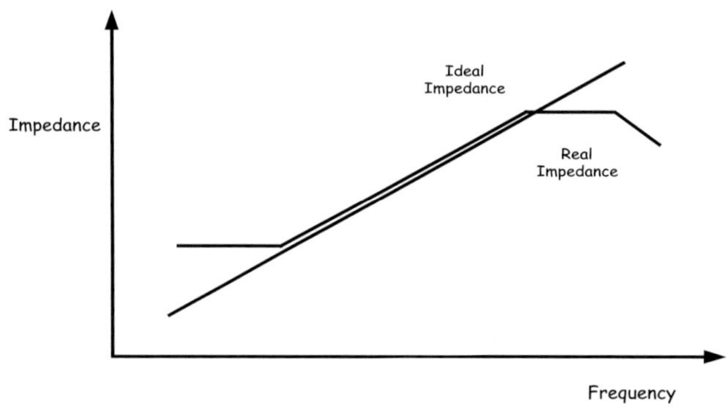

Figure 5.24 – Ideal and real (or non ideal) impedances at figure 5.23 grounded inductance simulator input

Using CCIIs, it is possible to implement a more useful inductance simulator, the floating one. From the GIC presented in figure 5.18, replacing Z_1 and Z_2 with two resistances and Z_3 with a capacitance, an equivalent inductance, valued R_1R_2C can obtained. The resulting circuit is pictured in figure 5.25 [2].

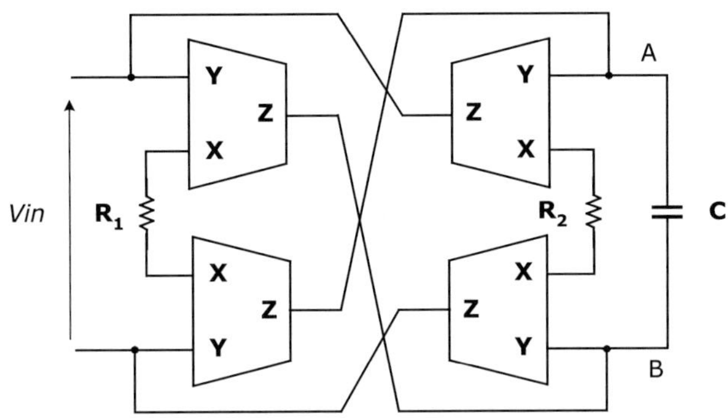

Figure 5.25 – CCII-based floating inductance simulator

The impedance seen between the two Y nodes is equal to sR_1R_2C if ideal components are considered. Taking into account the non ideal CCII model and assuming that all the CCIIs have the same value of C_Y, C_Z and R_Z, the impedance seen at the input of the circuit shown in figure 5.25 can be modelled by the scheme reported in figure 5.26 [6].

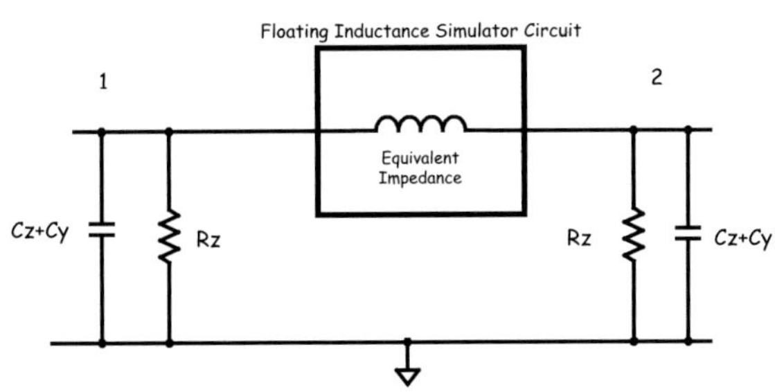

Figure 5.26 – Model of non ideal impedance seen between two input terminals in figure 5.25 circuit

Considering the real CCII model, the expected inductance value is $(R_1+2R_X)(R_2+2R_X)C$. It is important to notice that we have considered only a first order model, because parasitic impedances at capacitor C terminal have not been included.

The equivalent impedance has a frequency behaviour close to that given by the grounded inductance simulator (see figure 5.24), and similar considerations are possible about the percentage error, the quality factor and the operating frequency range. Anyway, this solution, suitable for integrated applications too, has the drawback that requires a floating capacitance between A and B nodes (figure 5.25). This problem can be overcome starting from the same circuit and grounding one of the terminals of Z_3 impedance. The result is the circuit pictured in figure 5.27 [6,8].

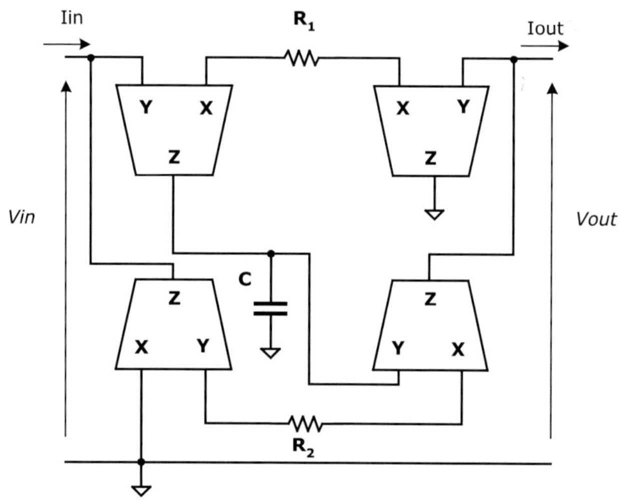

Figure 5.27 – Floating inductance simulator using grounded capacitor

Using this circuit, there is no need of implementing a floating capacitor, like in the previous case, and this is an important advantage in the fabrication of an equivalent inductance performed through fully integrated components.
Considering ideal CCIIs, an inductance valued R_1R_2C between nodes IN and OUT is seen. Taking into account CCII parasitic node impedances, the expected value for the equivalent impedance shown by the floating inductance simulator circuit is:

$$Z_{IN} = \frac{(R_1 + 2R_X)(R_2 + 2R_X)(1 + sR_Z(C_T + C))}{R_Z} \tag{5.24}$$

where C_T is equal to C_Y+C_Z. Also in this case, the total impedance graph is similar to that reported in figure 5.24 and the percentage relative error is lower than 2%. Previous considerations about Q factor and frequency range (see eq.s (5.22) and (5.23)) still hold.

Also in this case it is possible to replace the capacitor used in inductance simulation with a multiplier scheme because it is a grounded one. In fact, the proposed capacitance multiplier allows the simulation of grounded capacitors only. This error is not higher than the one obtained without capacitance multiplication, but a smaller frequency operating range, due to the capacitance multiplier non idealities, must be accepted.

The solution presented in figure 5.27 seems to be disadvantageous, because of the use of four CCIIs, but we must consider the fact that all the active components are equal, and this is a great advantage in integrated applications because the design can be done just for the single element.

Moreover, using dual output CCIIs, it is possible to obtain a floating equivalent inductance using only two active components, as reported in figure 5.28 [9].

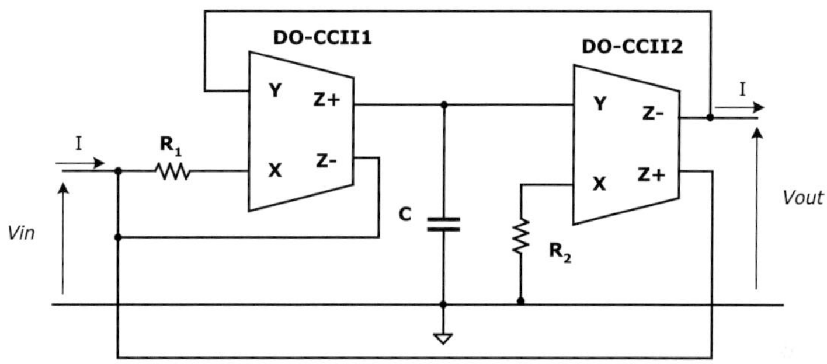

Figure 5.28 – DOCCII-based floating inductance simulator

In this case the equivalent inductance is still equal to R_1R_2C, as can be easily seen from figure 5.28 circuit.

This application of DOCCII confirms once more the fact that such a kind of current conveyors gives to analog designers a higher degree of freedom, with respect to op-amp solutions, allowing to implement more functions using less active elements.

Other inductance simulator topologies utilizing the evolved CCII blocks have been proposed in the literature. In this sense, the Current Controlled CCII (CCCII), introduced in the previous chapter, shows a parasitic resistance at X node that is electronically tunable. Using two CCCIIs it is possible to implement a grounded inductance simulator, as reported in figure 5.29 [10].

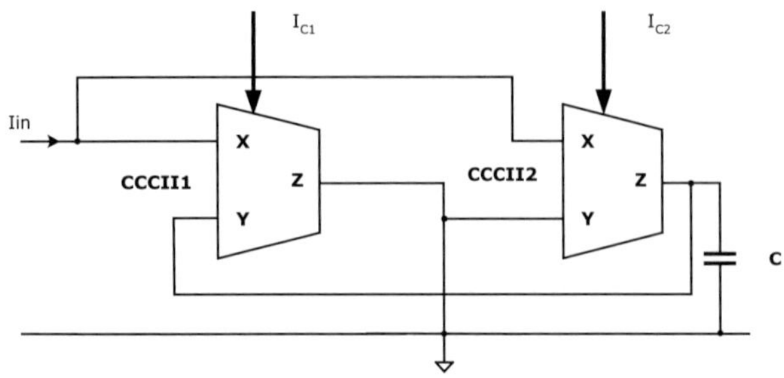

Figure 5.29 – CCCII-based grounded inductance simulator

As shown in the previous chapter (figure 4.8), the voltage V_X is given by:

$$V_X = R_X I_X + V_Y \tag{5.25}$$

Neglecting the CCCII non-idealities, it is possible to write the following expression for the equivalent input impedance:

$$Z_{IN} = sR_{X1}R_{X2}C \, // \, R_{X1} \, // \, R_{X2} \tag{5.26}$$

The input impedance Z_{IN} is given by the parallel of a grounded inductance and the two R_X resistances. The value of these two resistances affects the inductance value too, allowing an electronic control of the simulated inductance. The only drawback may be represented by the fact that the simulated impedance, even assuming ideal CCCIIs, is not a pure inductance. The range of the simulated inductance may be easily adjusted inserting, in series with both X nodes, a resistance, as pictured in figure 5.30 [11].

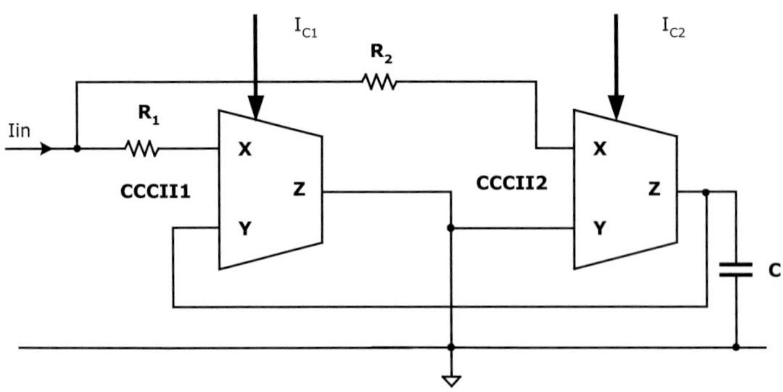

Figure 5.30 – Improved CCCII-based grounded inductance simulator

In this last case the equivalent impedance is the following :

$$Z_{IN} = s(R_1 + R_{X1})(R_2 + R_{X2})C//(R_1 + R_{X1})//(R_2 + R_{X2}) \tag{5.27}$$

Other similar configurations for the implementation of negative resistances and grounded or floating admittances have been proposed in [12,13,14,15].

5.3 CCII-BASED FILTERS

Like the op-amps, also CCIIs can be utilized in the design of analog filters, especially the current-mode ones. In the following, we will consider both passive and active filter implementation through current conveyors.
In the previous section a CCII-based capacitance multiplier has been introduced. The simplest application of this circuit, related to analog filters, is represented by a RC filter. Connecting a resistance to the input terminal of the capacitance multiplier previously described, as shown in figure 5.31, a low pass filter is obtained [16,17].

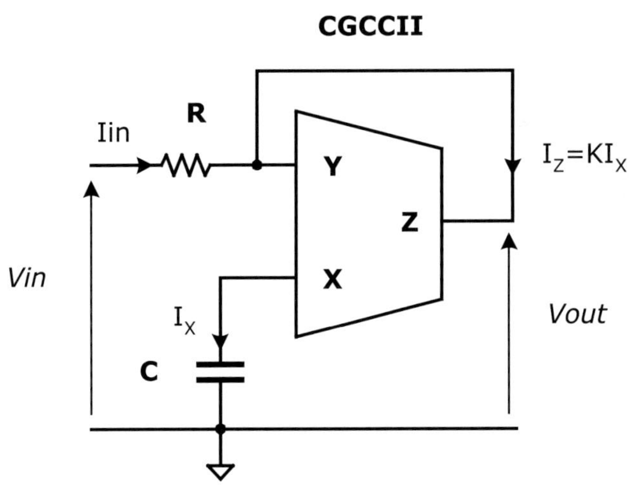

Figure 5.31 – CCII-based RC filter

Using a 100 KΩ resistance, a 50 pF capacitance and imposing a current gain of 50,000, it is possible to implement a low pass filter having very low cut-off frequencies (for example, lower than 1 Hz), as shown in figure 5.32. The advantage of this solution is represented by its intrinsic simplicity and by the fact that it is suitable for fully integrated implementation.

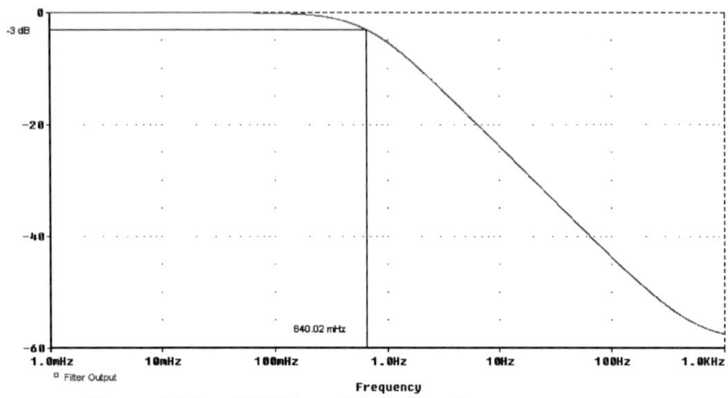

Figure 5.32 – CCII-based RC filter frequency response

The high valued inductances and capacitors, designed in the previous sections, can be employed in the implementation of a fully integrable LC band-pass filter, whose resonating frequency is equal, as well known, to :

$$f_R = \frac{1}{2\pi\sqrt{LC}} \tag{5.28}$$

From eq.(5.28), it comes that low resonating frequencies can be obtained only if high values for both capacitors and inductors are employed. In order to show how the proposed circuits can be applied for the implementation of a LC band pass filter, three different topologies, reported in figure 5.33 and indicated with A, B and C, have been implemented [6].

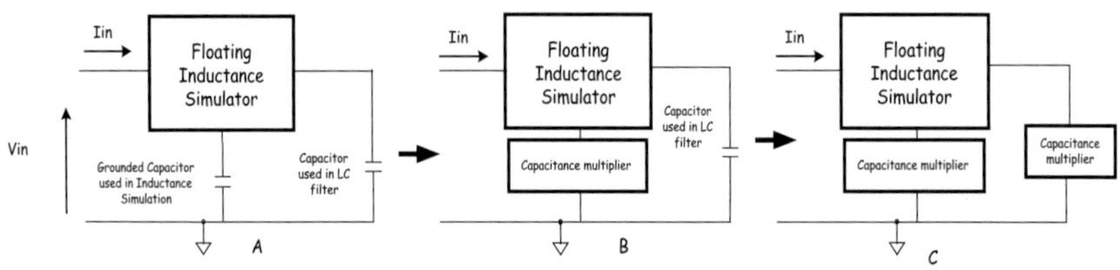

Figure 5.33 – CCII-based LC filter implementations

The filter topology A utilises a floating inductance simulator like that presented in figure 5.27 and two grounded capacitors. In the topology B the capacitor used in floating inductance simulation has been substituted with an equivalent capacitor obtained with a capacitance multiplier.
Topology C has been obtained by substituting also the capacitor used in the LC filter with another CGCCII-based capacitance multiplier. In this case the circuit is completely integrable on chip.
Three different values of resonance frequency have been considered. Simulation results, presented in table 5.1, show a good agreement with the theoretical values [6].

Band-Pass Filter Results					
Filter Topology	Inductance	Capacitance	Theoretical Resonance Frequency	Simulated Resonance Frequency	Error
A	20 H	20nF	251.65 Hz	255.62 Hz	1.57%
A	200 H	20nF	79.57 Hz	80.92 Hz	1.69%
A	2000 H	20nF	25.16 Hz	25.56 Hz	1.58%
B	20 H	20nF	251.65 Hz	254.65 Hz	1.19%
B	200 H	20nF	79.57 Hz	81.45 Hz	2.36%
B	2000 H	20nF	25.16 Hz	25.46 Hz	1.19%
C	20 H	20nF	251.65 Hz	243 Hz	3.43%
C	200 H	20nF	79.57 Hz	81.67 Hz	2.63%
C	2000 H	20nF	25.16 Hz	26.03 Hz	3.35%

Table 5.1 – Results from different filter topologies simulation and comparison with theoretical values

The capacitor used in the LC filter, equal to 20 nF, has been obtained from an integrable 20 pF capacitance and a multiplier with K factor equal to 1000. We can also note that the percentage error is slightly higher when all the capacitors needed for a LC filter are obtained by a capacitance multiplication.
In figure 5.34 the simulated and theoretical input currents of the equivalent LC band-pass filter are compared. It is possible to see that the LC filter response is like the expected one for a limited range of frequencies, and, at the resonance frequency, its Q value is lower than the theoretical one. However, in the other frequency ranges, the band-pass behaviour is ensured. The supply voltage used for the CCII used in the LC filter implementation is 1.5 V. The total power consumption is only 194 μW.

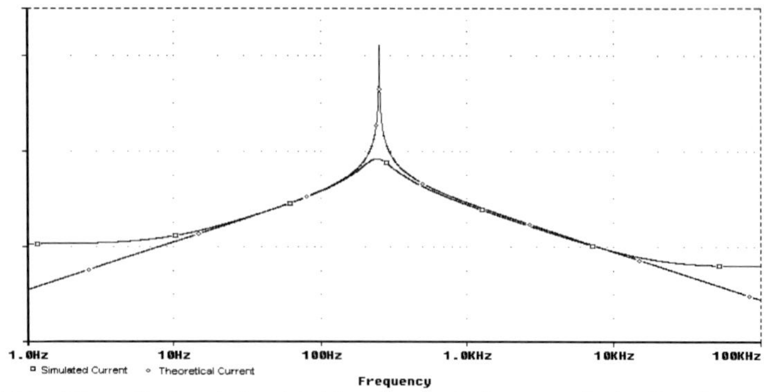

Figure 5.34 – Theoretical and simulated input currents for an equivalent band-pass filter

An equivalent LC filter can be implemented using grounded CCCII-based inductance simulators instead of floating ones.
In particular, simply adding a capacitance C_1 to the inductance simulator shown in figure 5.30, a second order band pass filter is easily derived, and is presented in figure 5.35 [11].

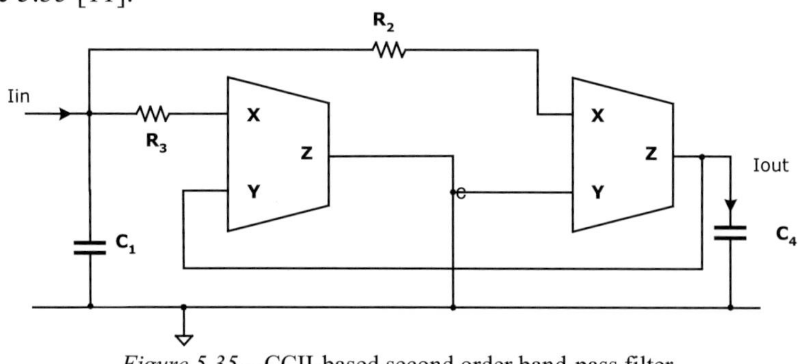

Figure 5.35 – CCII-based second order band-pass filter

Replacing CCIIs with CCCIIs, the filter shows an electronic control of the resonance frequency. In fact:

$$f_R = \frac{1}{2\pi\sqrt{LC}} \quad \text{where } C = C_1 \quad \text{and } L = (R_3 + R_{X1})(R_2 + R_{X2})C_4 \tag{5.29}$$

The CCCII biasing current modifies R_{Xi} values, allowing an electronic tuning of the band-pass filter. This solution presents some limitations, due to the fact that the output current flows into the capacitance used for the inductance simulation.

In order to overcome this problem, a second current output may be implemented in the second CCII shown in figure 5.35. In [11], a different solution that allows to increase Q factor is also proposed.
Another second order band-pass filter, having the current as both input and output signal, can be derived starting from the non ideal inductance simulator described in figure 5.29. The proposed circuit is reported in figure 5.36 [10].

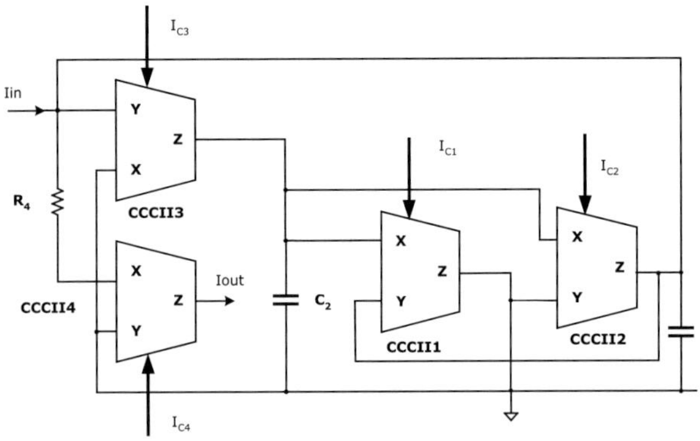

Figure 5.36 – CCCII-based grounded inductance simulator used in band-pass filter

In this figure, CCCII1 and CCCII2, together with capacitor C_1, implement the inductance simulator previously presented. CCCII3 operates as a controlled negative resistance, used to compensate the shunt resistance shown by the equivalent inductance. Finally, CCCII4 and resistance R_4 allow to perform a high output impedance. In this way, many identical cells may be directly and easily cascaded so to increase the filter order. The characteristic equations are the following:

$$\frac{1}{R_{EQ}} = \frac{1}{R_{X1}} + \frac{1}{R_{X2}} + \frac{1}{R_4 + R_{X4}} - \frac{1}{R_{X3}}$$

$$\omega_0 = \frac{1}{\sqrt{C_1 C_2 R_{X1} R_{X2}}} \qquad Q = \frac{R_{EQ}}{\sqrt{R_{X1} R_{X2}}}\sqrt{\frac{C_2}{C_1}} \qquad \text{Gain}(\omega_0) = \frac{R_{EQ}}{R_{X4} + R_4} \tag{5.30}$$

All the presented filters have been designed with a single output signal, so with them it is possible to obtain only an output function. In figure 5.37, it is shown how DOCCII may help in the implementation of filters able to perform simultaneously more than one filtering functions [18].

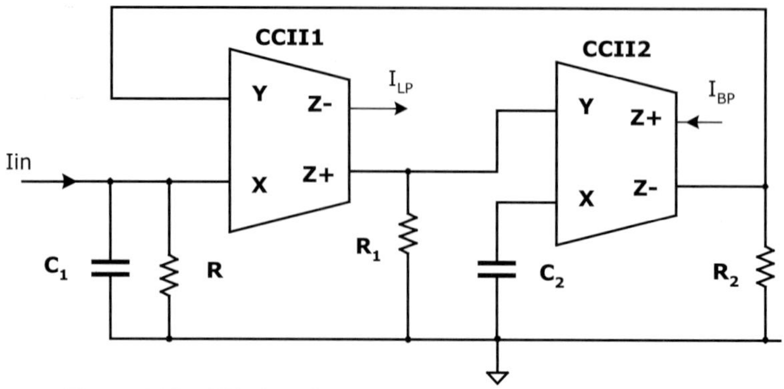

Figure 5.37 – CCII-based current mode low pass band pass filter

Two different functions, low-pass (LP) and band-pass (BP), have been obtained at the same time. The transfer functions are reported in the following formulas :

$$\frac{I_{LP}}{I_{IN}} = \frac{\frac{1}{C_1C_2R_1R_2}}{s^2 + \frac{s}{C_1R} + \frac{1}{C_1C_2R_1R_2}}$$

(5.31)

$$\frac{I_{BP}}{I_{IN}} = \frac{\frac{s}{C_1R_2}}{s^2 + \frac{s}{C_1R} + \frac{1}{C_1C_2R_1R_2}}$$

(5.32)

The centre frequency and the Q-factor values are given by the following equations :

$$\omega_0 = \frac{1}{\sqrt{C_1C_2R_1R_2}} \qquad Q = R\sqrt{\frac{C_1}{C_2R_1R_2}}$$

(5.33)

A good characteristic shown by this solution is represented by the fact that these two parameters are tuneable separately. The centre frequency gain is easily controlled, because it is equal to the ratio between R and R_2 resistances. In fact :

$$\text{GAIN}(\omega_0) = \frac{R}{R_2}$$

(5.34)

From figure 5.37 it appears clear how CCII and its evolutions are really flexible and versatile blocks. Using only two active blocks and few passive elements a very versatile current mode filter has been obtained. It must be noted that all passive components are grounded ones and this is a very important feature for integrated applications.
Following this philosophy a high number of different solutions have been presented in literature [19,20,21,22,23,24,25,26,27,28,29,30,31,32,33], some of which will be analysed in this book. In figure 5.38 a current mode filtering cell that presents three different outputs is reported [21].

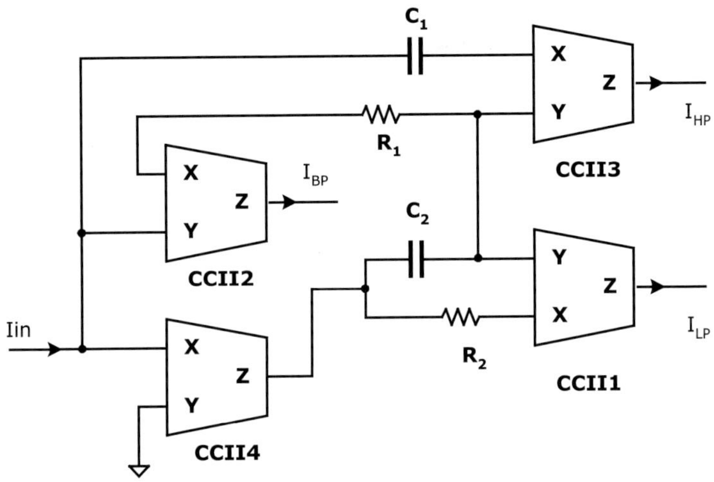

Figure 5.38 –Single-input multiple-output CCII-based current mode filter

Starting from a single input current and using four CCIIs, this filter allows to have low-pass, band-pass and high-pass output functions accordingly to the following expressions:

$$\frac{I_{LP}}{I_{IN}} = \frac{\frac{1}{C_1C_2R_1R_2}}{D(s)}, \qquad \frac{I_{BP}}{I_{IN}} = \frac{\frac{s}{C_1R_1}}{D(s)}, \qquad \frac{I_{HP}}{I_{IN}} = \frac{-s^2}{D(s)}$$

$$D(s) = s^2 + \frac{s}{R_1C_1} + \frac{1}{R_1C_1R_2C_2} \tag{5.35}$$

The resonating frequency and the Q-factor are given by :

$$\omega_0 = \frac{1}{\sqrt{C_1 C_2 R_1 R_2}} \qquad Q = \sqrt{\frac{R_1 C_1}{R_2 C_2}} \tag{5.36}$$

The topology proposed in figure 5.38 shows all floating passive components. The same solution which has only grounded passive elements is shown in figure 5.39 [27]. It must be pointed out that an extra resistance has been added and that the CCIIs have been replaced by DOCCIIs, so confirming once more the importance of the dual output configuration [23,24,25,26,27]. Moreover, another filtering function is implemented, in fact, together with low-pass, band-pass and high-pass outputs, an output current performing the notch function, and indicated with I_{01}, has been implemented (see eq.(5.37)).

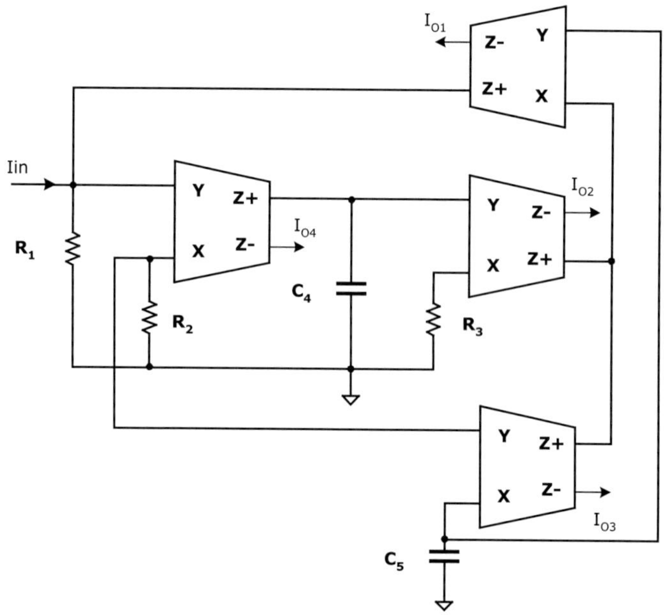

Figure 5.39 – Single-input multiple-output DOCCII-based current mode filter

$$\begin{cases} \dfrac{I_{O1}}{I_{IN}} = \dfrac{G_2G_3 + s^2C_4C_5}{s^2C_4C_5 + sG_1C_4 + G_2G_3} & \text{NOTCH} \\ \dfrac{I_{O2}}{I_{IN}} = -\dfrac{G_2G_3}{s^2C_4C_5 + sG_1C_4 + G_2G_3} & \text{LOWPASS} \\ \dfrac{I_{O3}}{I_{IN}} = -\dfrac{s^2C_4C_5}{s^2C_4C_5 + sG_1C_4 + G_2G_3} & \text{HIGHPASS} \\ \dfrac{I_{O4}}{I_{IN}} = -\dfrac{sG_2C_4}{s^2C_4C_5 + sG_1C_4 + G_2G_3} & \text{BANDPASS} \end{cases} \tag{5.37}$$

$$\omega_0 = \frac{1}{\sqrt{C_4C_5R_2R_3}} \qquad Q = R_1\sqrt{\frac{C_5}{R_2R_3C_4}} \tag{5.38}$$

After the considerations done, the design guidelines in the CCII-based filters can be summarized as follows:

1. Lowering the number of active components.
2. Lowering the number of passive components.
3. Using grounded passive components.
4. Performing more than one transfer function.
5. Making tunable the characteristics of the designed filter (possibly, by means of a single parameter variation).

In figure 5.40 a good example of such a kind of approach is reported [31,32].

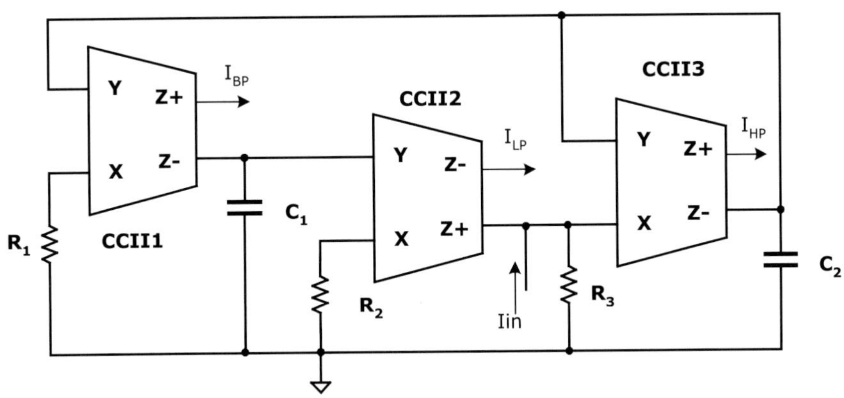

Figure 5.40 –DO-CCII-based current-mode biquad filter

This filter is a current-mode biquad, allowing to obtain simultaneously three functions: low-pass, band-pass and high-pass. The centre frequency and the Q factor are adjustable independently, as shown in eq.(5.39).

$$\omega_0 = \frac{1}{\sqrt{C_1 C_2 R_1 R_2}} \qquad Q = R_3 \sqrt{\frac{C_2}{R_1 R_2 C_1}} \tag{5.39}$$

The oscillating frequency can be set varying R_1 and/or R_2, as, for example, in figure 5.41, where these two resistances have been taken equal.

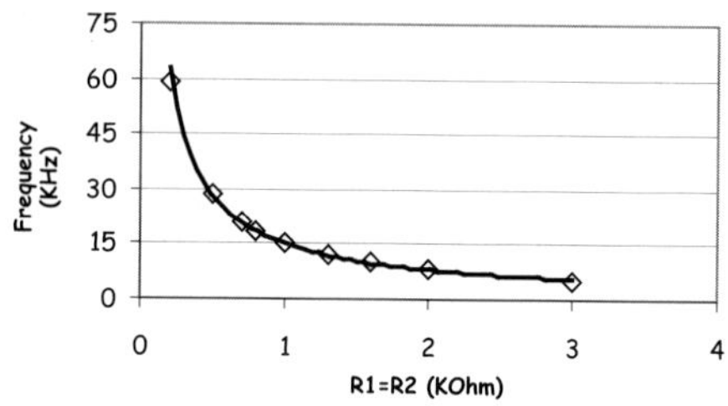

Figure 5.41 – Oscillating frequency vs. R1=R2

On the contrary, different values of R_3 do not affect, as shown in figure 5.42, the oscillating frequency, which is, in fact, imposed by the other passive elements.

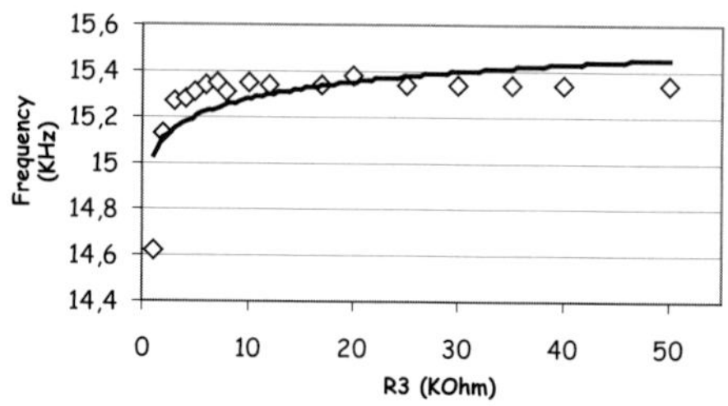

Figure 5.42 – Oscillating frequency vs. R3

An increase of R_1 and/or R_2 resistances lowers the oscillating frequency (see figure 5.41) but also the filter Q-factor (accordingly to eq.(5.39) and also as shown in figure 5.43).

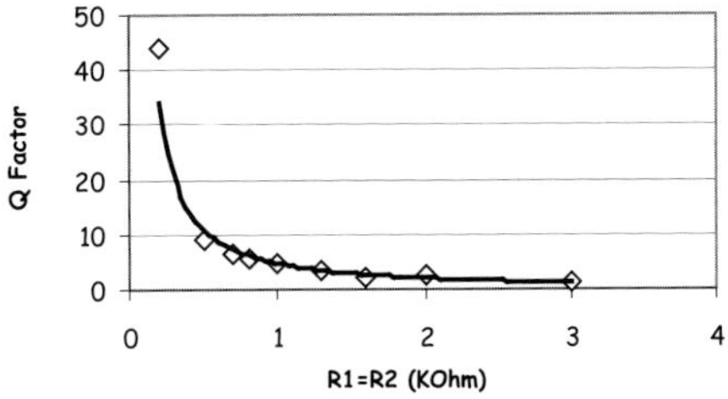

Figure 5.43 – Q-factor vs. R1=R2

If this reduction of Q-factor value is not acceptable, the same factor can be independently increased modifying R_3 value, as reported in figure 5.44.

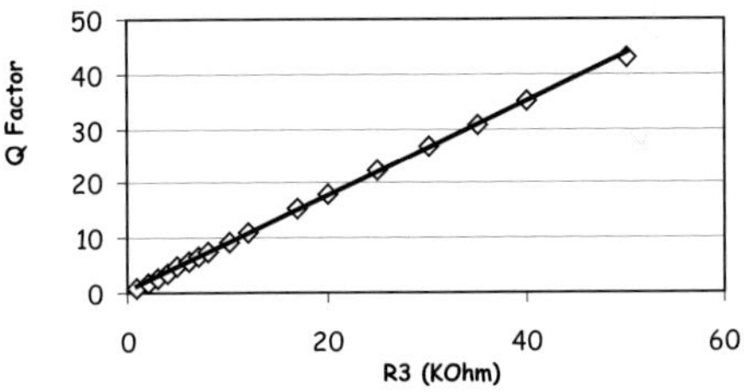

Figure 5.44 – Q-factor vs. R3

All these degrees of freedom are possible thanks to the fact that there is an independent control of the oscillating frequency and the Q-factor value. These feature represents a great improvement in filter design.

In figure 5.45 the simulated frequency responses have been presented. It appears clear that higher Q values may show an unwanted behaviour in the low-pass and high-pass response, so a compromise has to be accepted between a flat response of low and high pass functions and a high Q value for the band-pass one.

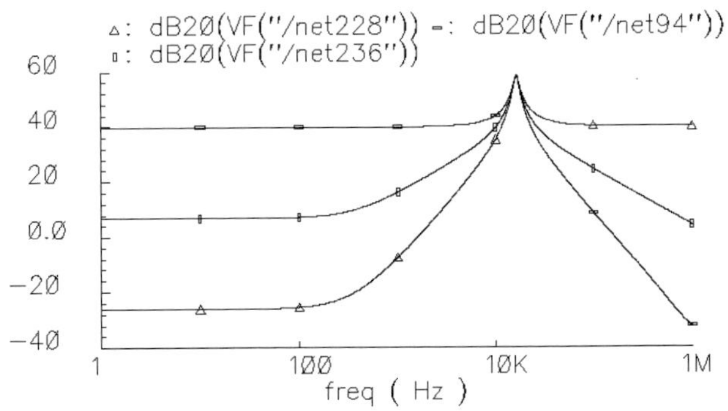

Figure 5.45 –DO-CCII based biquad filter frequency response

CCII-based filters can be designed following a different approach, that leads to consider a multiple input and single output topology. An example is presented in figure 5.46 [24].

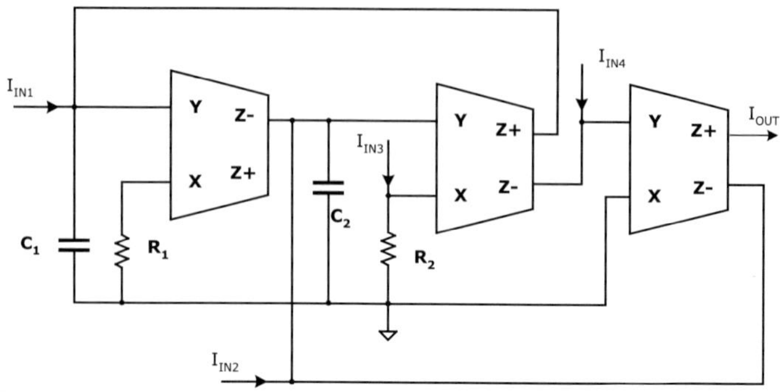

Figure 5.46 –Single-input multiple output DO-CCII-based universal filter

There are four different inputs, named I_{IN1}-I_{IN4}, and a single output. The filter has been built up using integrators, so a lower sensitivity with respect to active and passive components has been obtained.

All the passive components are grounded, and four different output functions can be obtained: low-pass, band-pass, high-pass and notch.

The overall transfer function is given by eq. (5.40).

$$I_{OUT} = \frac{-s^2 I_{IN3} + \left(\dfrac{s}{C_2R_2}\right)I_{IN2} - \left(\dfrac{1}{C_1C_2R_1R_2}\right)I_{IN1} - \left(s^2 + \dfrac{1}{C_1C_2R_1R_2}\right)I_{IN4}}{s^2 + \dfrac{s}{C_2R_2} + \dfrac{1}{C_1C_2R_1R_2}} \tag{5.40}$$

Modifying input current values, all the functions presented above can be obtained:

$$\begin{cases} I_{IN1} = I_{IN2} = I_{IN4} = 0, & \text{Input Current Signal } I_{IN3} \Rightarrow \text{HIGHPASS} \\ I_{IN1} = I_{IN3} = I_{IN4} = 0, & \text{Input Current Signal } I_{IN2} \Rightarrow \text{BANDPASS} \\ I_{IN2} = I_{IN2} = I_{IN4} = 0, & \text{Input Current Signal } I_{IN1} \Rightarrow \text{LOWPASS} \\ I_{IN1} = I_{IN2} = I_{IN3} = 0, & \text{Input Current Signal } I_{IN4} \Rightarrow \text{NOTCH} \end{cases} \tag{5.41}$$

It has to be noted that there are not particular matching condition to be satisfied in order to implement the filter response. From eq. (5.40) it is possible to write the centre frequency and the quality factor as follows:

$$\omega_0 = \frac{1}{\sqrt{C_1 C_2 R_1 R_2}} \qquad Q = \sqrt{\frac{R_2 C_2}{R_1 C_1}} \tag{5.42}$$

One of the most important characteristics of a filter is the possibility to control easily its response. Multi-function filters allow to implement several different transfer functions. Basically, the designer gives separate expressions for resonating frequency and bandwidth (or Q factor). A good filter allows us to modify these two characteristics separately. As seen in eq.(5.42), this is not completely true for the circuit shown in figure 5.46. Accepting the fact to use more active and passive components, several solutions have been presented in the literature.
In figure 5.47 an excellent current-mode universal filter is reported [33].

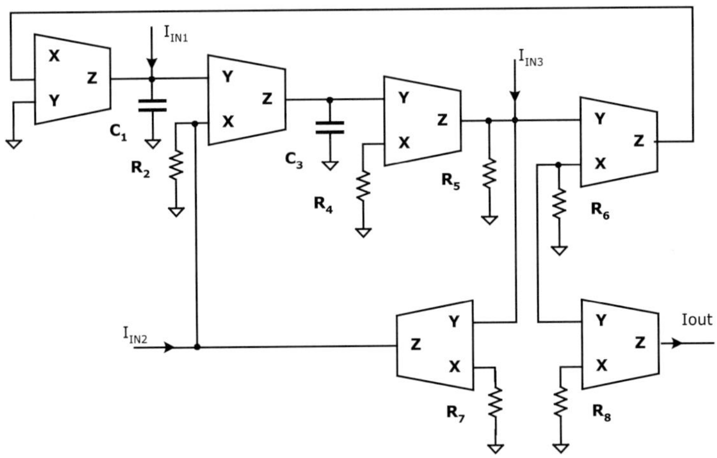

Figure 5.47 –CCII-based universal current-mode filter

Performing the routine analysis the output current expression is given by :

$$I_{OUT} = -G_8 \frac{-s^2 C_1 C_3 I_{IN3} - sC_1 G_4 I_{IN2} + G_2 G_4 I_{IN1}}{s^2 C_1 C_3 G_5 + sC_1 G_4 G_7 + G_2 G_4 G_6} \tag{5.43}$$

Settings different values for the four input currents, many transfer functions are obtained as follows:

$$\begin{cases} I_{IN2} = I_{IN3} = 0 \Rightarrow \text{LOWPASS} \\ I_{IN1} = I_{IN2} = 0 \Rightarrow \text{HIGHPASS} \\ I_{IN1} = I_{IN3} = 0 \Rightarrow \text{BANDPASS} \\ I_{IN2} = 0, I_{IN1} = I_{IN3}, G_5 = G_6 \Rightarrow \text{NOTCH} \end{cases} \tag{5.44}$$

Modifying the current input and imposing some limitation to the impedance values allows to implement other functions, such as low-pass and high-pass notch filters. Resonance frequency and quality factor are respectively:

$$\omega_0 = \sqrt{\frac{R_5}{C_1 C_3 R_2 R_4 R_6}} \qquad Q = \sqrt{\frac{R_7}{C_1 R_2 R_6}} \tag{5.45}$$

From the previous equations, it is clear that the resonance frequency can be adjusted by controlling the grounded resistance R_4 without affecting the bandwidth. The bandwidth is controlled by R_7 without changing ω_0. Moreover, it is also possible to control, by means of R_8 grounded resistance, the current gain, without effects on resonance frequency and/or bandwidth. In fact, the current gains for the low-pass, band-pass and high-pass functions are the following:

$$\begin{cases} G = \frac{R_6}{R_8} \Rightarrow \text{LOWPASS} \\ G = \frac{R_5}{R_8} \Rightarrow \text{HIGHPASS} \\ G = \frac{R_7}{R_8} \Rightarrow \text{BANDPASS} \end{cases} \tag{5.46}$$

Current conveyors can be successfully employed in the implementation of voltage mode applications too. A possible example of such a use of CCIIs is represented by the voltage mode multifunction filter presented in figure 5.48 [25].

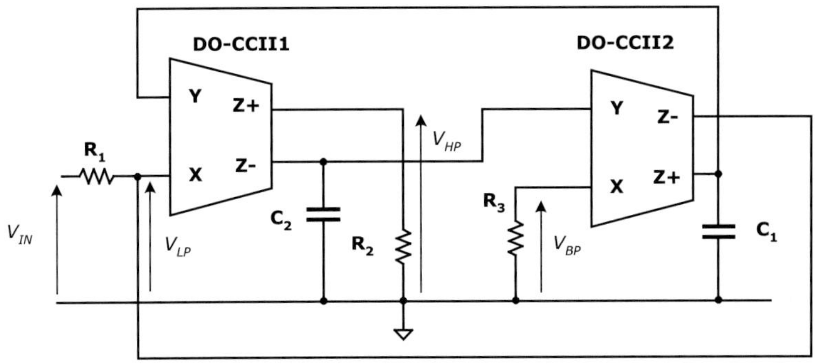

Figure 5.48 –DO-CCII-based voltage-mode filter

The circuit shown in figure 5.48 presents a single voltage input V_{IN} and three voltage outputs, performing low-pass, high-pass and band-pass functions, as follows:

$$\begin{cases} \dfrac{V_{LP}}{V_{IN}} = \dfrac{G_1G_2G_3}{s^2C_1C_2G_2 + sC_1G_2G_3 + G_1G_2G_3} \\ \dfrac{V_{BP}}{V_{IN}} = \dfrac{sC_1G_1G_2}{s^2C_1C_2G_2 + sC_1G_2G_3 + G_1G_2G_3} \\ \dfrac{V_{HP}}{V_{IN}} = \dfrac{-s^2C_1C_2G_1}{s^2C_1C_2G_2 + sC_1G_2G_3 + G_1G_2G_3} \end{cases} \tag{5.47}$$

The resonance frequency and the quality factor Q, as reported in eq.(5.48), are adjustable by changing the grounded resistance R_3 and the floating resistance R_1. In fact:

$$\omega_0 = \frac{1}{\sqrt{C_1C_2R_1R_3}} \qquad Q = \sqrt{\frac{R_3C_2}{R_1C_1}} \tag{5.48}$$

In conclusion, we have presented a number of CCII-based filters, which differs among them from the possibility of performing voltage and current multi-filtering operations. Their LV LP characteristics are linked to the performance of the conveyors which have been utilised in their internal topologies.

5.4 OSCILLATORS.

Current conveyors can be also successfully used in the design of oscillators. In fact, they have shown to provide wider bandwidth and a best accuracy with respect to the traditional op-amps. Moreover, they do not present a serious disadvantage of op-amp based oscillators. In fact, it has been proved [34] that the limitation of the finite GBW product affects the frequency oscillation as well as the oscillation condition. For this reason, CCII-based oscillators have been proposed in the literature since more than 20 years ago [35,36,37].

Anyway, in signal processing a wide number of voltage-mode solutions have been developed. That's why it is not unusual to find transformation methods that allow to obtain a current-mode solution from its voltage-mode counterpart [38,39,40,41]. An example of this approach is constituted by the design of a Wien oscillator, whose solution using a VCVS as amplifier block in a Wien bridge is shown in figure 5.49 [41].

The design equations for this oscillator are the following:

$$\text{Oscillation Frequency } \omega_0 = \frac{1}{\sqrt{C_1 C_2 R_1 R_2}}; \quad \text{Oscillation Condition } K = 1 + \frac{R_1}{R_2} + \frac{C_2}{C_1} \tag{5.49}$$

being K the VCVS voltage gain.

We can consider R_1 as a part of the active block; in this case the VCVS is replaced by a VCCS, whose gain is K/R_1. Using different solutions for the VCCS implementation, different expression of the gain have to be considered.

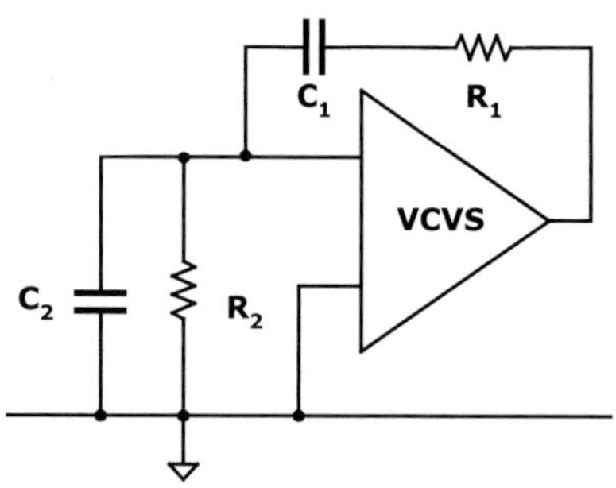

Figure 5.49 –Wien Oscillator scheme

CCIIs can be employed in the design of any kind of active block, then other CCII-based Wien oscillator configurations can be designed (figure 5.50) [41]. Oscillation frequency and condition are the same, provided that the following relationships are verified:

$$\text{Topology(a)}: \quad R_A = \frac{R_1}{K}$$
$$\text{Topology(b)}: \quad R_A = \frac{R_1}{K-1} \tag{5.50}$$

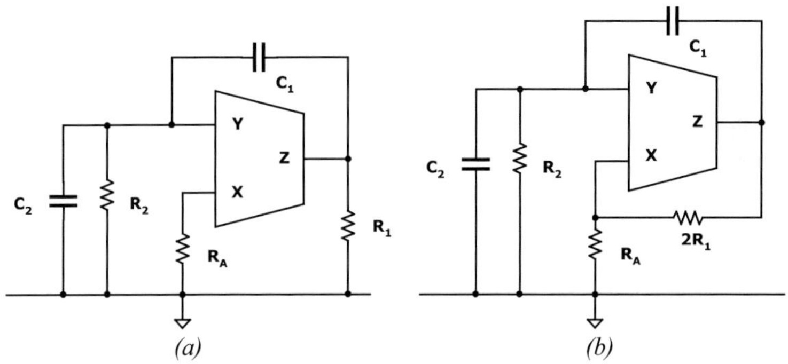

Figure 5.50 – CCII-based Wien oscillators

Two other CCII-based solutions for Wien sinusoidal generators are reported in figure 5.51, where the following equations have to be verified:

$$\text{Topology(a)}: \quad R_A = \frac{R_1}{K}$$
$$\text{Topology(b)}: \quad R_A = \frac{R_1 R_2}{K R_2 - R_1} \tag{5.51}$$

The topologies shown in figure 5.50(a) and 5.51(a) are single-frequency oscillators, while in Figure 5.51(b) we have reported a variable-frequency oscillator whose oscillation frequency is dependent on only one resistance value (R_2) [41].

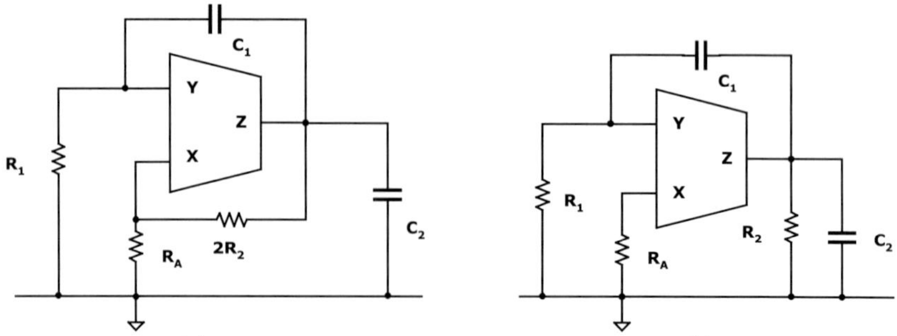

Figure 5.51 – Two others CCII-based Wien oscillators

Another CCII-based Wien oscillator has been proposed in [42] and is reported in figure 5.52.
The design equations for this oscillator are the following :

$$\text{Oscillation Frequency } \omega_0 = \sqrt{\frac{1+\frac{R_2}{2R_3}}{C_1C_2R_1R_2}}\text{; Oscillation Condition } R_4 = R_1 + 2R_3\left(\frac{R_1}{R_2}+\frac{C_2}{C_1}\right) \tag{5.52}$$

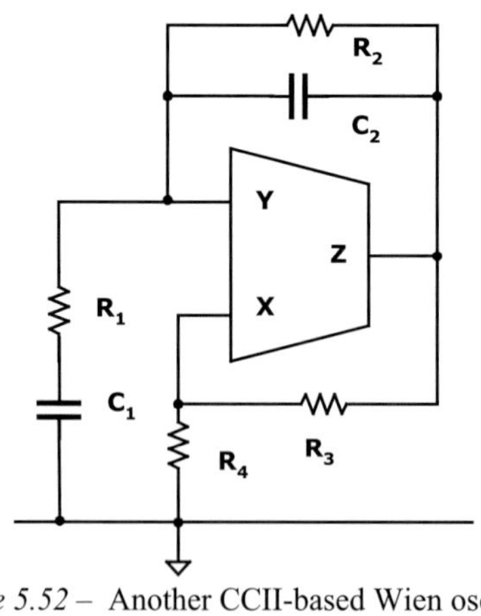

Figure 5.52 – Another CCII-based Wien oscillator

In order to have all the capacitances grounded, the circuit has been modified in that shown in figure 5.53 [42].
The design equations for this oscillator are the following:

$$\text{Oscillation Frequency } \omega_0 = \frac{1}{C_1R_1}; \text{ Oscillation Conditions } C_1R_1 = C_2R_2;\ 2\frac{R_3}{R_4} = \frac{R_2}{2R_1} - 1 \tag{5.53}$$

Using the current conveyor topology presented in figure 2.21, some PSPICE simulations have been performed. The oscillation frequency has been varied modifying the capacitances value. For values down to 20 pF and taking into account the effect of some parasitic impedances, the difference between expected and simulated values is lower than 2%. In the same conditions the THD has a value always lower than 3%. Moreover it has to be considered the fact that without changing the CCII topology it has been possible to set oscillation frequencies from 1 KHz up to 1.3 MHz. For capacitance values lower than 20 pF the effect of the parasitic capacitances increases, and the same occurs to the difference between theory and simulations. This is only due to the fact that all the parasitic impedances should be considered, leading to a extremely complicated theoretical analysis. It has to be considered too that, according to eq.(5.53), lower capacitances mean higher oscillation frequencies, where the transistor model is little bit more complicated. Anyway the simulations have shown a good agreement with theory and the CCII-based oscillator has shown good performance too.

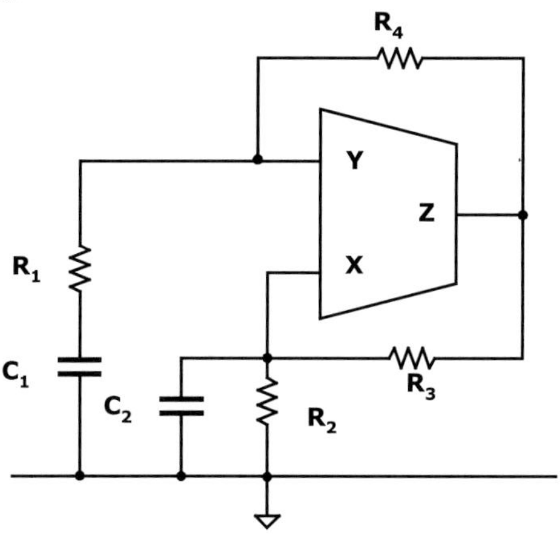

Figure 5.53 – CCII-based Wien oscillator using all the capacitances grounded

It is highly desirable to have a CCII-based oscillator with all the passive components grounded. In this sense, a possible solution is represented by the circuit in figure 5.54, where three basic CCIIs have to be used [43].

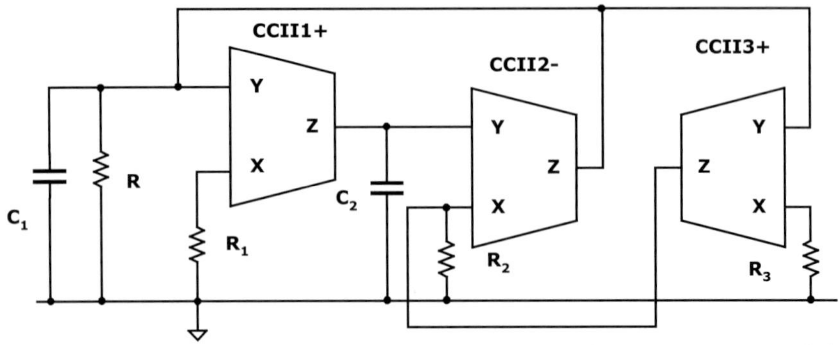

Figure 5.54 – CCII-based oscillator using all the passive components grounded

The design equations for this oscillator are the following:

$$\text{Oscillation Frequency} \quad \omega_0 = \frac{1}{\sqrt{C_1 C_2 R_1 R_2}}; \quad \text{Oscillation Condition } R_3 = R \tag{5.54}$$

This oscillator has the advantage that a grounded resistor (R or R_3) can be utilised to control the oscillation condition without affecting the frequency of oscillation, since R_1 or R_2 can control the oscillation frequency independently from the oscillation condition.

Other topologies of CCII-based oscillators show some particular features, like that proposed in [44]; in particular, the use of only positive CCIIs, the use of four (or two) grounded resistors and two (or four) grounded capacitors, and the possibility of having separate oscillation frequency and condition (figure 5.55). The design equations for the oscillator shown in figure 5.55 are the following:

$$\text{Oscillation Frequency} \quad \omega_0 = \frac{1}{\sqrt{C_3 R_4 R_2 C_5}}; \quad \text{Oscillation Condition } C_1 = C_2 \tag{5.55}$$

Another possible oscillator can be implemented starting from dual output CCIIs and only grounded resistances and capacitances [45]. In this case only two CCII blocks are needed (figure 5.56).

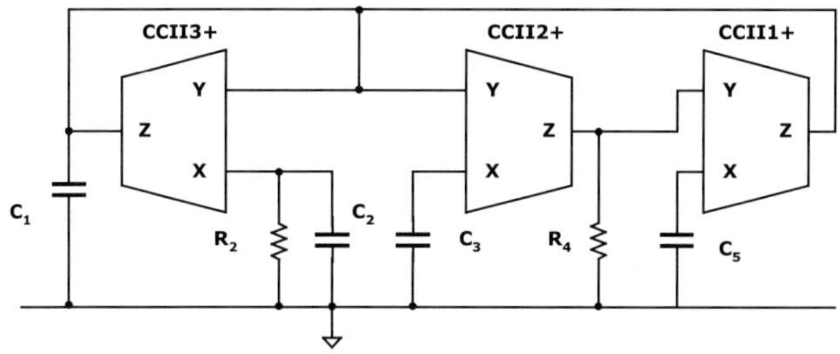

Figure 5.55 – All positive CCII-based oscillator

The design equations for this oscillators are the following :

$$\text{Oscillation Frequency } \omega_0 = \frac{1}{\sqrt{C_1C_2R_1R_2}}; \quad \text{Oscillation Condition } R_1 = R \tag{5.56}$$

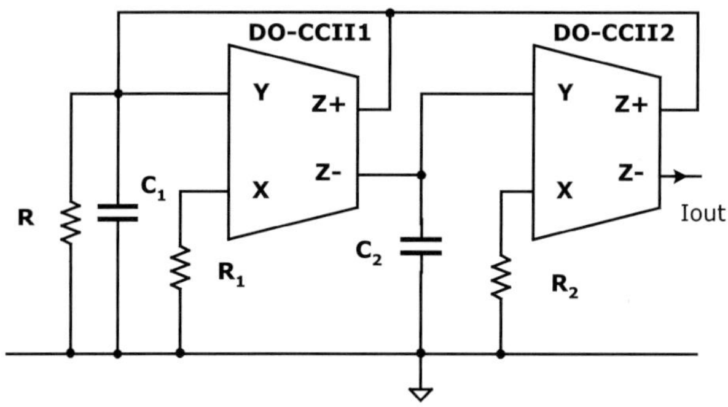

Figure 5.56 – Dual output CCII based oscillator

A circuit implementing the same oscillating frequency of that shown in eq.(5.56) but using basic CCII blocks is reported in figure 5.57 [46]. In this case R_3 controls the oscillation condition (R_3=R_2) without affecting the oscillation frequency.

Using different active blocks, interesting solutions can be implemented with particular features like the multiphase property [47], where the oscillator can produce a number of output currents equally spaced in phase. The basic circuit is shown in figure 5.58.

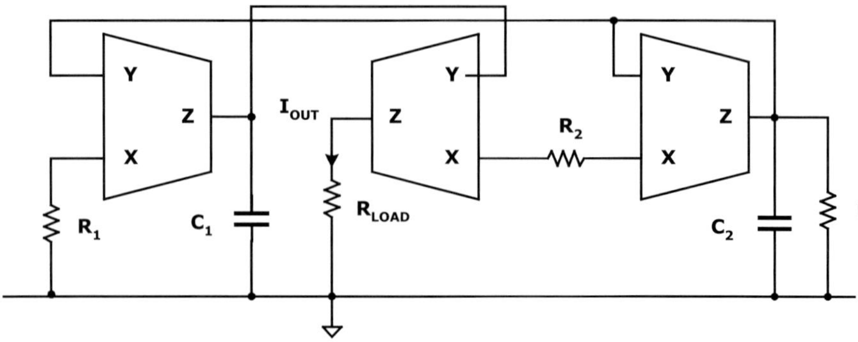

Figure 5.57 – Another CCII-based oscillator

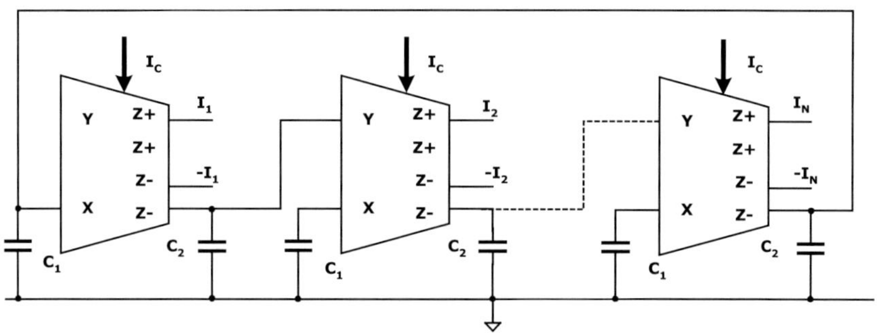

Figure 5.58 – Multiphase sinusoidal oscillator based on current conveyors

All the passive components are grounded and no external resistances are used, so the circuit seems particularly attractive for integrated applications. Moreover, the frequency of oscillation can be controlled independently from the oscillation condition.

Other simple current-mode oscillators can be implemented using both first and second generation current conveyors [48]. An example is shown in figure 5.59. For C_1=C_2=C, the oscillation frequency is equal to:

$$\omega_0 = \frac{1}{C\sqrt{R_1 R_2}} \qquad (5.57)$$

It possible to set this frequency by controlling the value of R_1.

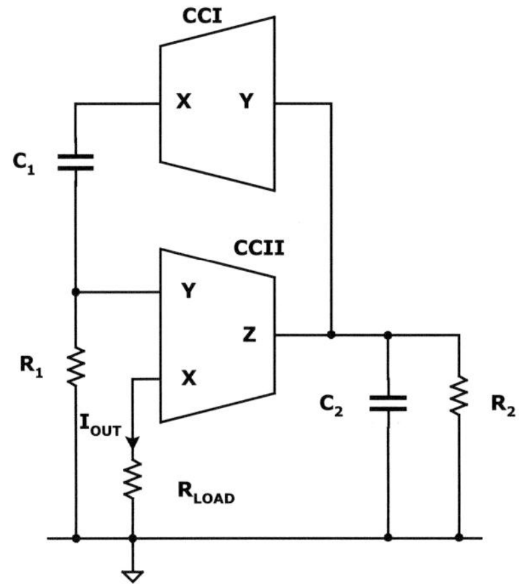

Figure 5.59 – Oscillator based on first and second generation current conveyors

To prevent saturation current, that can cause high power consumption and distortion, the passive resistance R_{LOAD} have been replaced by an active non-linear resistance. This circuit, shown in figure 5.60, is based on four transistor diode-connected (M_1–M_4) and two equivalent current sources (M_5, M_6).

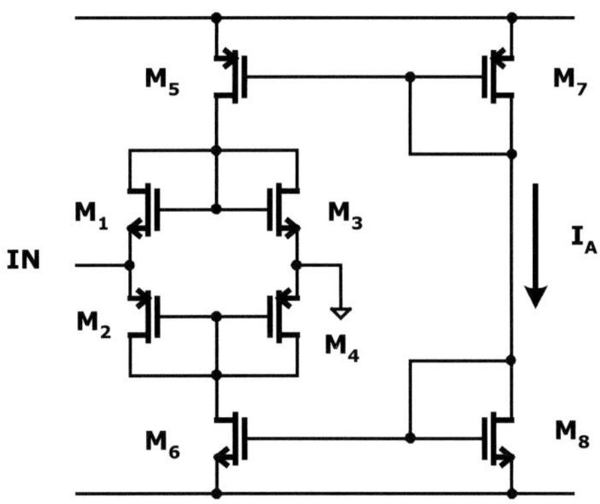

Figure 5.60 – Equivalent topology implementing a grounded resistance

Setting the DC bias current I_A it is possible to control the equivalent input resistance at node IN. In fact, at small voltage amplitudes, the equivalent resistance at node IN is given by $1/\sqrt{4\beta I_0}$, where $\beta = \mu_N C_{OX} W_N / L_N = \mu_P C_{OX} W_P / L_P$, μ is the carrier mobility, W/L the transistor geometrical ratio, C_{OX} the oxide capacitance per unit area and I_0 is the DC transistor biasing current.

Figure 5.61 shows, for R_2=10 kΩ , $C_1 = C_2$ = 2pF, the simulated and theoretical variation of the oscillation frequency as a function of R_1 (corresponding total harmonic distortions (THD) in % are also indicated for various frequencies). There is an excellent agreement between theoretical and simulated results. The THD, for a corresponding output peak-to-peak amplitude of 50 mV, is lower than 1.5 % and is mainly caused by the external non-linear resistance R_A and the parasitic non-linear resistance R_{XY} of CCI [48].

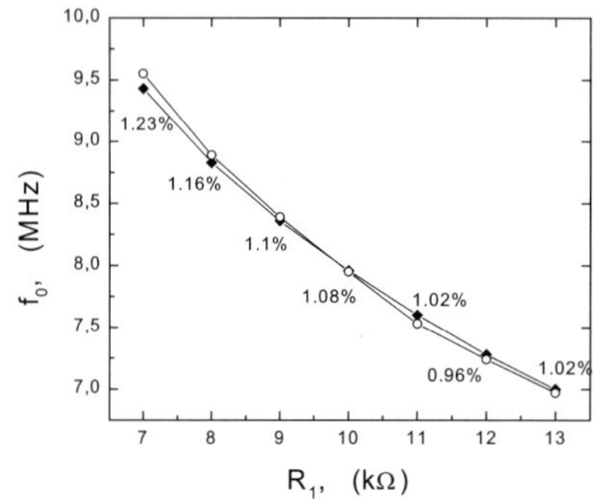

Figure 5.61 –Theoretical and simulated frequencies as a function of R_1
♦ Simulated O Theoretical

Finally, we propose two topologies of oscillators using only one active building block. The first one, shown in figure 5.62, implements an oscillator based on DVCCII [49].

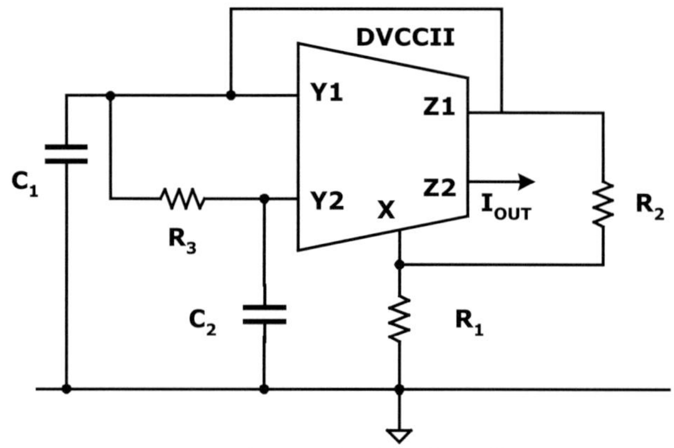

Figure 5.62 – DVCCII-based oscillator

The design equations for this oscillator are the following :

$$\text{Oscillation Frequency } \omega_0 = \sqrt{\frac{2}{C_1 C_2 R_2 R_3}}; \text{ Oscillation Condition } R_1 < \frac{R_3 C_2}{C_1 + C_2} \tag{5.58}$$

The second configuration is based on a FDCCII and is reported in figure 5.63 [50]. The output node Z1 has been doubled, so to obtain the output current on a node not involved in the oscillator operation.

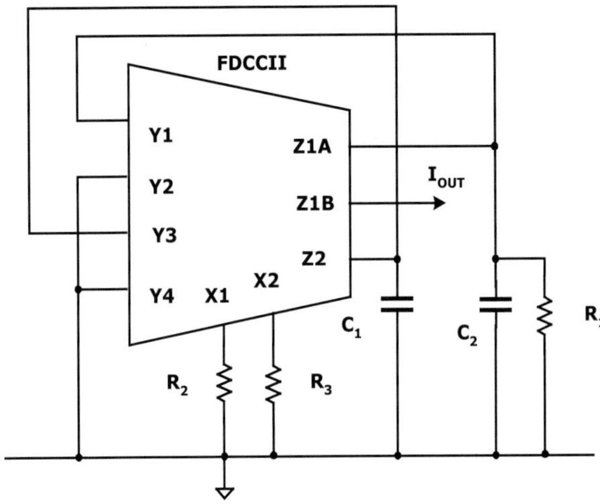

Figure 5.63 – FDCCII-based oscillator

With respect to the solution previously described, this case seems more attractive since all the passive components are grounded.
The design equations for this oscillator are the following:

$$\text{Oscillation Frequency } \omega_0 = \frac{1}{\sqrt{C_1 C_2 R_2 R_3}}; \quad \text{Oscillation Condition } R_1 = R_3 \tag{5.59}$$

As for the filters, also for what concerns the oscillators, we have proposed some CCII-based solutions. In our opinion, the best topologies are those which allows an independent control of the oscillation frequency and condition. Also in this case the oscillator LV LP characteristics depend directly on the utilised CCII performance.

5.5 CCII APPLICATIONS

Current conveyors, being an alternative to the classical operational amplifier, can be employed with success in an extremely wide range of different applications. In chapter 4 we have shown how the basic CCII can be used as building block in designing more complicated conveyors topologies. In this section we present some applications of current conveyors.

5.5.1 Current conveyor with reduced parasitic impedance

One of the basic characteristics of a current conveyor is represented by its X node impedance. Ideally this impedance should be zero, but in non-ideal circuits its non-zero value has to be taken into account. In the previous sections we have shown how this parasitic impedance gives an error in the designed analog functions, even if this problem can be overcome by avoiding loads whose value is comparable to that of the parasitic impedance.
A possible solution to this problem can be represented by the circuit in fig. 5.64, where two positive current conveyors are employed to obtain a negative current conveyor, highlighted into the dotted lines, so showing a low parasitic impedance at X node [51].

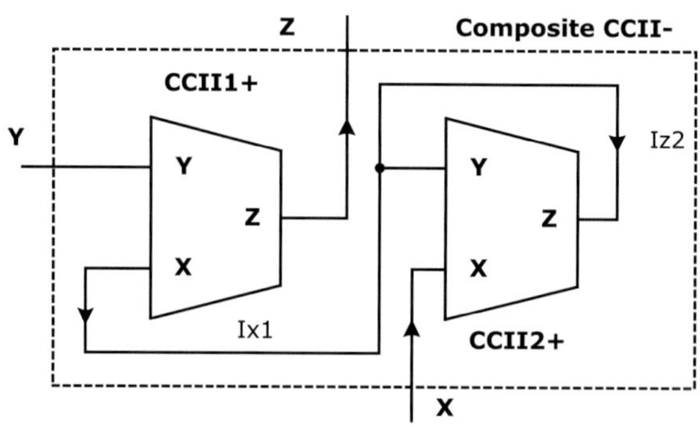

Figure 5.64 – Composite CCII with low X node parasitic impedance

In fact, when a voltage is applied to the Y node of the first CCII and a load is connected to the X node of the second CCII, it follows that $I_{x1} = I_{z2} = -\beta_2 I_{x2}$. Then, the parasitic equivalent impedance between X and Y is equal to $R_x(1-\beta_2)$ and heavily decreases since β_2 tends to 1.

5.5.2 DCCII-based four quadrant multiplier

The DCCII, introduced in the previous chapter 4, is an "evolution" of the basic CCII towards the differential applications. As stated before, it can be thought as composed by basic current conveyors and resistances. A very interesting application of the DCCII is represented by a transconductance four quadrant multiplier, shown in figure 5.65 [52], where transistors M1 and M2 operate in the ohmic region. The two output currents are given by :

$$I_{z1} = K(V_{G1} - V_{G2})(V_1 - V_2) ;$$
$$I_{z2} = K(V_{G1} - V_{G2})(V_2 - V_1) \tag{5.60}$$

so showing a proportionality with the voltage difference applied at the transistor gates and that between V1 and V2.

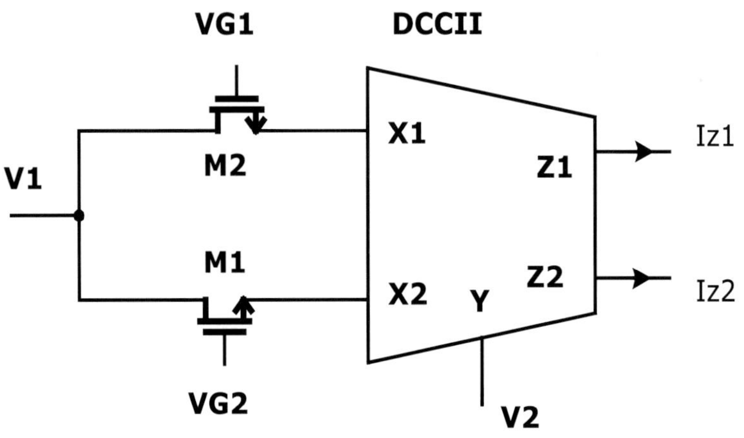

Figure 5.65 – A four quadrant multiplier using the DCCII

5.5.3 Sensor interface

Figure 5.66 shows an interface circuit for piezoresistive pressure sensors based on CCIIs [53]. The advantage of this circuit in the sensor interface is the capability of offset compensation and a versatile current-mode configuration with current output and current or voltage input. The output voltage is linearly proportional to the piezo variation.
It is assumed that Rp+ is the piezoresistor, whose variation is Rp+=Rpo(1+x), being Rpo the resistance at reference pressure. A current source configuration is obtained through CCII1+. The current Iz1 is equal to Ix1 and therefore is set by V_{REF} and R_1 values. The change in piezo value affects the voltage at Z1 node as follows:

$$V_{z1} = \frac{V_{REF}}{R_1} Rpo(1 + x) + V_{OFF} \tag{5.61}$$

The output current of CCII2+ also follows the input current, since Y3 is a high impedance node, while the output current of CCII3+ is given by the ratio between V_{Z1} and R_2. The two currents I_{Z2} and I_{Z3} are then added to obtain the output current Io and consequently:

$$V_o = (I_{Z3} - I_{Z2})R_3 . \tag{5.62}$$

In this manner, a linear dependence between Io and the piezo variation x is obtained and can be expressed as follows:

$$Io = I_{Z3} = I_{Z2} = \frac{V_{Z1}}{R_2} - \frac{V_{REF}}{R_1} = \frac{x\ Rpo\ V_{REF}}{R_1 R_2} + k \tag{5.63}$$

where k is a constant output offset value which can be also set to zero by suitable choice of V_{REF}, V_{OFF}, R_1 and R_2.

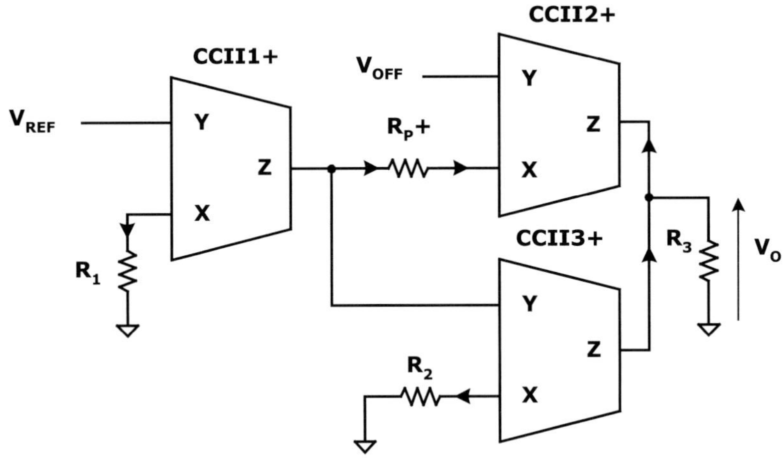

Figure 5.66 – Interface circuit for piezoresistive pressure sensors based on CCIIs

Other interfaces based on current conveyors can be developed using suitable CCII topologies having good performance in terms of low noise, low offset, and capable to operate with low voltage low power characteristics.

In conclusion, in this chapter a wide number of analog applications of second generation current conveyors and their evolutions have been proposed. These applications cover exhaustively the same amount of circuits which generally can be implemented using operational amplifiers (from basic applications such as voltage and current amplifiers, integrator and differentiators, to impedance simulators, filters and oscillators), confirming once more time the importance of this device as a basic building block, especially for LV LP applications.

References.

[1] A. S. Sedra G. W. Roberts. "Current conveyor theory and practice". In *Analogue IC design: The current mode approach*. Peter Peregrinus, London, 1990.

[2] C. Toumazou, A. Payne, D. Haigh. *Analogue IC design: The current mode approach*. Peter Peregrinus, London, 1990.

[3] G. A. Rincon-Mora. Active capacitor multiplier in Miller-compensated circuits. IEEE Transactions on Solid-State Circuits. nr. 1; vol. 35; 2000; pp. 26-32.

[4] G.Di Cataldo, G.Ferri, S.Pennisi. Active capacitance multipliers using current conveyors, Proceedings of International Symposium of Circuits and Systems, 1998; Monterey, USA.

[5] G.Ferri, S.Pennisi. A 1.5 V Current-Mode Capacitance Multiplier, Proceedings of International Conference on Microelectronics, 1998; Monastir, Tunisia.

[6] G.Ferri, N.Guerrini. High valued passive element simulation using low-voltage low-power current conveyors for fully integrated applications. IEEE Transactions on Circuits and Systems II. n.4; vol.48; 2001; pp. 405-409.

[7] A. Sedra, K. Smith, *Microelectronic circuits,* Oxford University Press, 1998.

[8] W. Kiranon, P. Pawarangkoon. Floating inductance simulation based on current conveyors. Electronics Letters. nr. 21; vol. 33; 1997; pp. 1748-1749.

[9] P. V. Ananda Mohan. Grounded capacitor based grounded and floating inductance simulation using current conveyors. Electronics Letters. nr. 11; vol. 34; 1998; pp. 1037-1038.

[10] A. Fabre, O. Saaid, F. Wiest, C. Boucheron. High-frequency high-Q BiCMOS current-mode bandpass filter and mobile communication application. IEEE Journal of Solid-State Circuits. nr. 4; vol. 33; 1998; pp. 614-624.

[11] A.Fabre, H.Amrani, O.Saaid. Current-mode bandpass filters with Q magnification. IEEE Transactions on Circuit and Systems-II. nr. 12; vol. 43; 1996; pp. 839-842.

[12] S. Ozoguz, On the realization of floating immitance functions using CCIIs. Proceedings of the European Conference Circuit, Theory and Design; 1997; Budapest.

[13] O. Çiçekoglu, H. H. Kuntman. Single CCII+ based active simulation of grounded inductors. Proceedings of European Conference Circuit, Theory and Design; 1997; Budapest.

[14] M. T. Abuelma'atti, N. A. Tassadduq. New negative immittance function simulators using current conveyors. Microelectronics Journal. nr. 30; 1999; pp. 911-915.

[15] M. T. Abuelma'atti, N. A. Tassadduq. Electronically tunable capacitance multiplier and frequency-dependent negative-resistance simulator using the current controlled current conveyor. Microelectronics Journal. nr. 30; 1999; pp. 869-873.

[16] G.Ferri, S.Pennisi, S.Sperandii. A low voltage CMOS 1-Hz low pass filter, Proceedings of International Conference on Electronics, Circuits and Systems, 1999; Cyprus.

[17] P.De Laurentiis, G.Ferri, G.Palumbo, S.Pennisi. A low-pass 1-Hz 2V-supply current-conveyor based filter. Proceedings European Circuits and Systems Conference, 1999. Bratislava, Slovakia.

[18] H. O. Elwan, A. M. Soliman. A novel CMOS current conveyor realisation with an electronically tunable current mode filter suitable for VLSI. IEEE Transactions on Circuit and Systems-II. nr. 9; vol. 43; 1996; pp. 663-670.

[19] F. Seguin, A. Fabre. New second generation current conveyor with reduced parasitic resistance and bandpass filter application. IEEE Transactions on Circuit and Systems-I. nr. 6; vol. 48; 2001; pp. 781-785.

[20] C. M. Chang. Multifunction biquadratic filters using current conveyors. IEEE Transactions on Circuit and Systems-II. nr. 11; vol. 44; 1997; pp. 956-958.

[21] S. Ozoguz, A. Toker, E. O. Ciçekoglu. New current-mode universal filters using only four (CCII+)s. Microelectronics Journal. nr. 30; 1999; pp. 255-258.

[22] C. A. Karybakas, C. A. Papazoglou. Low sensitive CCII-based biquadratic filters offering electronic frequency shifting. IEEE Transactions on Circuit and Systems-II. nr. 5; vol. 46; 1999; pp. 527-538.

[23] O. Ciçekoglu. Current–mode biquad with a minimum number of passive elements. IEEE Transactions on Circuit and Systems-II. nr. 1; vol. 48; 2001; pp. 221-222.

[24] E. O. Gunes, A. Toker, S. Ozoguz. Insensitive current-mode universal filter with minimum components using dual output current conveyors. Electronics Letters. nr. 7; vol. 35; 1999; pp. 524-525.

[25] C. M. Chang, M. J. Lee. Voltage-mode multifunction filter with single input and three outputs using two compound current conveyors. IEEE Transactions on Circuit and Systems-I. nr. 11; vol. 46; 1999; pp. 1364-1365.

[26] M. T. Abuelma'atti, N. A. Tassadduq. A novel single-input multiple output current-mode current-controlled universal filter. Microelectronics Journal. nr. 29; 1998; pp. 901-905.

[27] O. Ciçekoglu. High output impedance current-mode four function filter with reduced number of active and passive elements using dual output current conveyor. Analog Integrated Circuits and Signal Processing. nr. 28; 2001; pp. 201-204.

[28] S. Ozoguz, A. Toker, O. Ciçekoglu. First-order allpass sections-based current-mode universal filter using ICCIIs. Electronics Letters. nr. 17; vol. 36; 2000; pp. 1443-1444.

[29] A. M. Soliman. Generation of current conveyor-based all-pass filters from op amp-based circuits. IEEE Transactions on Circuit and Systems-II. nr. 4; vol. 44; 1997; pp. 324-330.

[30] S. Ozoguz, A. Toker, E. O. Gunes. New high-Q band-pass filter configuration using current controlled current conveyor based all-pass filters. Proceedings of the IEEE International Conference on Electronic Circuits and Systems, 2001; Malta.

[31] M. C. Piccirilli. Current conveyor based universal biquad filter. Proceedings of the IEEE International Conference on Electronic Circuits and Systems, 2001; Malta.

[32] G.Ferri, N.Guerrini, M.C.Piccirilli. Low voltage current conveyor-based universal biquad filter. Proceedings of the IEEE International Conference on Industrial Electronics, 2002; L'Aquila, Italy.

[33] M. T. Abuelma'atti, N. A. Tassadduq. A novel three-input and one output universal current-mode filter using plus-type CCIIs. Microelectronics Journal. nr. 30; 1998; pp. 287-292.

[34] A.Budak. *Passive and active network analysis and synthesis*. Houghton Mifflin, Boston, 1974.

[35] A.M. Soliman. Simple sinusoidal active RC oscillators. International .Journal of Electronics. vol.39; 1975; pp.455-458.

[36] A.M. Soliman. A novel variable frequency sinusoidal oscillator using a single current conveyor. Proc. IEE 1978; vol.66, pp.800.

[37] A.M. Soliman. Realization of frequency dependent negative resistance circuits using two capacitors and a single current conveyor. Proc. IEE 1978; vol.125, pp.1336-1337.

[38] G.Roberts, A.S.Sedra. All current-mode frequency selective circuits. Electronics Letters. nr.12; vol.25; 1989; pp.759-760.

[39] G.Roberts, A.S.Sedra. A general class of current amplifier-based biquadratic filter circuits. IEEE Transactions on Circuits and Systems, nr.4; vol.39; 1992; pp.759-760.

[40] A.Carlosena, L.Serrano, S.Porta. Current-mode multiple feedback filters. IEEE Transactions on Circuits and Systems. nr.2; vol.40; 1993; pp.141-143.

[41] P.A.Martinez, S.Celma, I.Gutierrez. Wien-type oscillators using CCII+. Analog Integrated Circuits and Signal Processing. vol.7; 1995; pp.139-147.

[42] A. M. Soliman, A. S. Elwakil. Wien oscillators using current conveyors. Computers and Electrical Engineering. nr. 25; 1999; pp. 45-55.

[43] A. M. Soliman. Synthesis of grounded capacitor and grounded resistor oscillators. Journal of the Franklin Institute. nr. 336; 1999; pp. 735-746.

[44] M. T. Abuelma'atti, H. Al-Daghrier. New single element controlled sinusoidal oscillator employing CCII+. Microelectronics Journal. nr. 29; 1998; pp. 83-86.

[45] A. M. Soliman. Current mode CCII oscillators using grounded capacitors and resistors. International Journal of Circuit Theory and Applications. nr. 26; 1998; pp. 431-438.

[46] A. M. Soliman. New grounded-capacitor current-mode oscillators using single-output CCIIs. Journal of Circuit, Systems and Computers. nr. 3; vol. 8; 1998; pp. 363-378.

[47] M. T. Abuelma'atti, M. A. Al-Qahtani. A new current-controlled multiphase sinusoidal oscillator using translinear current conveyors. IEEE Transactions on Circuits and Systems. nr.7; vol.45; 1998; pp.881-885.

[48] H.Barthelemy, G.Ferri, N.Guerrini. A 1.5 V CCII-based tunable oscillator for portable industrial applications. Proceedings of the IEEE International Conference on Industrial Electronics, 2002, L'Aquila, Italy.

[49] S. S. Gupta, R. Senani. Grounded-capacitor current-mode SRCO: novel application of DVCCC. Electronics Letters. nr. 3; vol. 36; 2000; pp. 195-196.

[50] C. M. Chang, B. M. Al-Hashimi, H. P. Chen, S. H. Tu, J. A. Wan. Current mode single resistance controlled oscillators using only grounded passive components. Electronics Letters. nr. 19; vol. 38; 2002; pp. 1071-1072.

[51] A.Fabre, H.Barthelemy. Composite second-generation current conveyor with reduce parasitic resistance. Electronics Letters. nr. 5; vol. 30; 1994; pp. 377-278.

[52] H. O. Elwan, A. M. Soliman. CMOS differential current conveyors and applications for analog VLSI. Analog Integrated Circuits and Signal Processing. nr. 11; 1996; pp. 35-45.

[53] J.Samitier, M.Puig-Vidal, S.Bota, C.Rubio, S.Siskos, T.Laopoulos. A current-mode interface circuit for a piezoresistive pressure sensor. IEEE Transactions on Instrumentation and Measurements. nr. 3; vol. 47; 1998; pp. 708-709.

Appendix A

In [A1] it is proved how any linear, non reciprocal network can be represented starting from simple passive and active elements.
Passive elements needed for this representation are resistors and capacitors, while the active ones are two-terminal basic networks, introduced in this appendix.
The active components are named *nullator* and *norator* and are shown in figure A.1, with their main features.

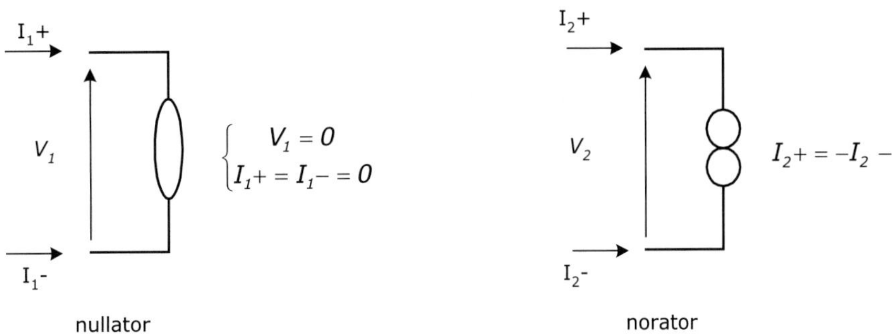

Figure A.1 – Nullator and norator

The *nullator* is characterized by zero voltage and current, while in the *norator* the current flowing into a node is equal to the current coming out from the other node. Voltage and current magnitudes are not defined.
Starting from these networks and considering the theory developed by Tellegen [A2], in 1964, Carlin introduced an ideal amplifier which utilises *nullators* and *norators,* naming the obtained two-port network as *nullor* [A1]. In fact, as stated before, since any active network can be represented in terms of *nullators* and *norators*, the simplest component is their combination and the nullor can be considered as the natural ideal active block.

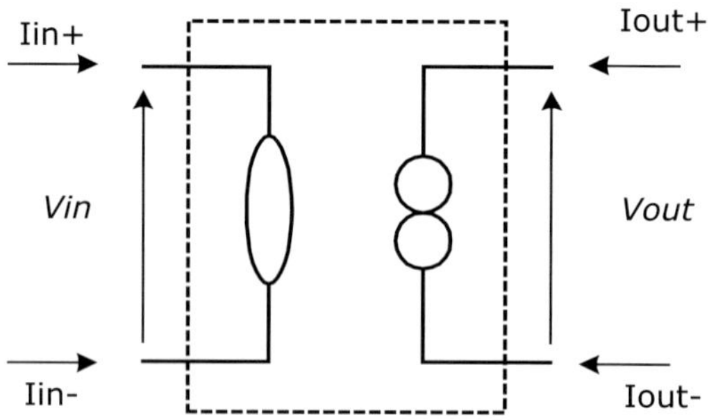

Figure A.2 – Nullor as a combination of nullator and norator

Nullor satisfies Tellegen's requirements for an ideal amplifier, which are zero input voltage and current, and arbitrary output voltage and current. As a consequence, the device has an infinite voltage gain and has become the starting point of every approach to amplifier theory. From nullor properties it comes that input and output impedance levels are not defined. Moreover, even if this ideal device is built-up, it will be impossible to use it in analog networks without feedback. Starting from these considerations, Tellegen [A1] proposed the four possible basic configurations, reported in figures 1.25 and 1.26, which cover all the possible combinations through the use of four kind of amplifiers (VCVS or voltage amplifier, VCCS or transconductance amplifier, CCVS or transresistance amplifier and CCCS or current amplifier).
It is interesting to show how OAs and negative CCIIs are easily derived from the primitive concept of *nullor*. In fact, grounding a *norator* terminal of the *nullor,* OA is obtained with its well-known characteristics (see fig. A.3) [A3].

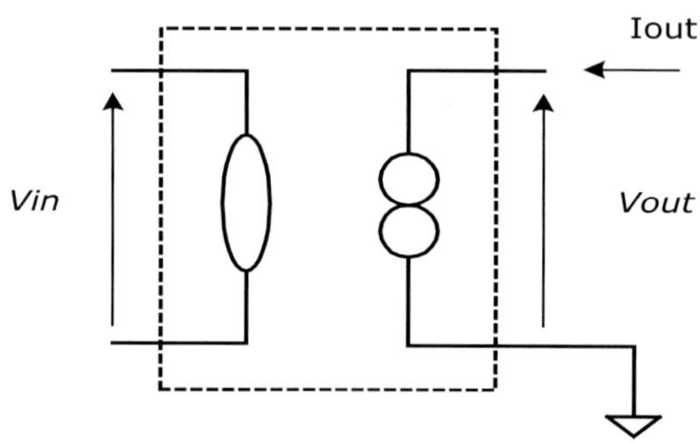

Figure A.3 – OA obtained from nullor

Furthermore, connecting one *nullator* to one *norator* terminal, we obtain a possible simple model for CCII-, as shown in fig. A.4 [A3].

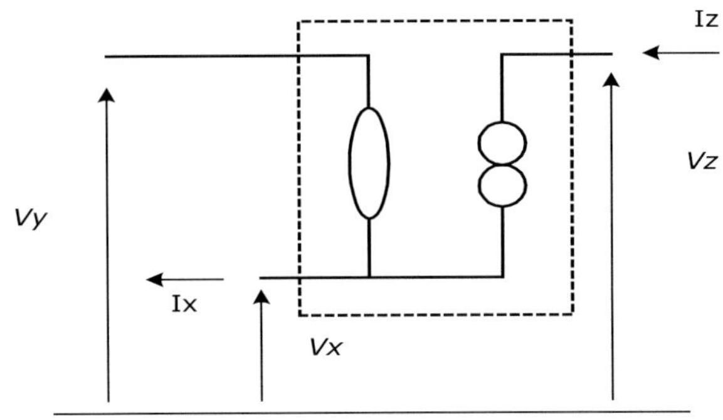

Figure A.4 – CCII- obtained from nullor

These considerations suggest that OAs and negative CCIIs, easily derived from the basic ideal block (the nullor), theorised by Carlin, can be considered as primitive active elements in analog design.

References.

[A1] H. J. Carlin. Singular network elements. IEEE Transactions on Circuit Theory. CT-11; 1964; pp. 67-72.

[A2] B. H. Tellegen. La recherche pour une serie complete d'elements de circuit ideaux non-lineaires. Rendiconti del seminario matematico e fisico di Milano 1954; Milano.

[A3] R. Carbeza and A. Carlosena, Analog universal active Device: Theory, design and applications. Analog Integrated Circuits and Signal Processing. nr. 12; 1997; pp. 153-168.

Appendix B

In this Appendix, experimental results concerning measurements on some integrated current conveyors will be presented. In particular, we have considered five of the presented topologies, fabricated in a 0.35 μm CMOS technology, provided by Austria Mikro Systems. The circuits have been designed without particular requirements, except for what concerns the X node parasitic impedance. That's why the results found in the previous chapters cannot be used for a comparison and new simulations using Spectre have been performed.
The first fabricated CCII topology is reported again in figure B.1, while in figure B.2 the microphotograph of the chip is presented.

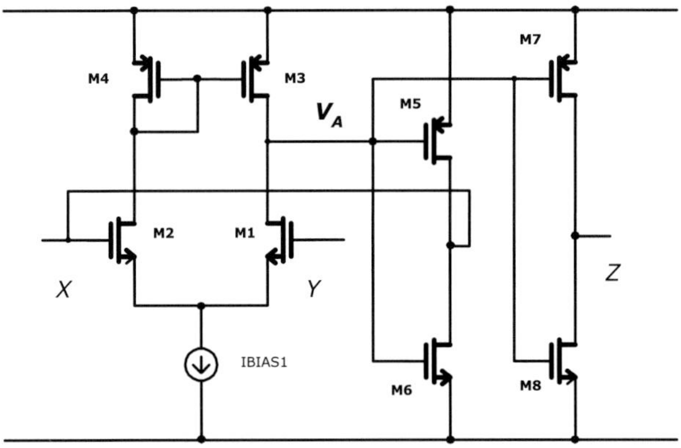

Figure B.1 – First CCII topology fabricated

Figure B.2 – Microphotograph of figure B.1 topology

First of all, the bandwidth of the current conveyor has been evaluated. A sinusoidal signal having an amplitude of 100 mV has been applied to the Y node, while X node has been left unloaded. The frequency of the signal has been varied and the corresponding signal amplitude at X node has been measured. The results concerning the frequency response have been summarised in the graph shown in figure B.3.

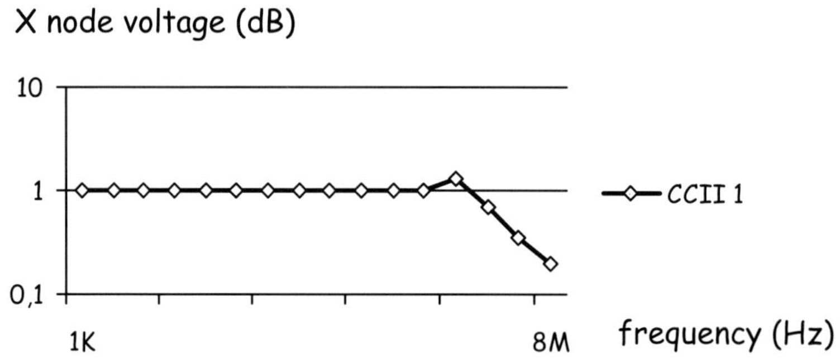

Figure B.3 – Measured frequency response of figure B.1 topology

The frequency response shows, in the experimental measures, a peak at about 1 MHz which did not result from simulations. Another interesting parameter is represented by the dynamic range, which has evaluated applying a constant voltage to Y node and measuring the corresponding voltage at X node. In other words, we have performed a group of measurements very similar to a classical DC sweep simulation. The experimental data (V_X vs. V_Y) are reported in figure B.4. The dynamic range is quite large and in a very good agreement with the simulation results.

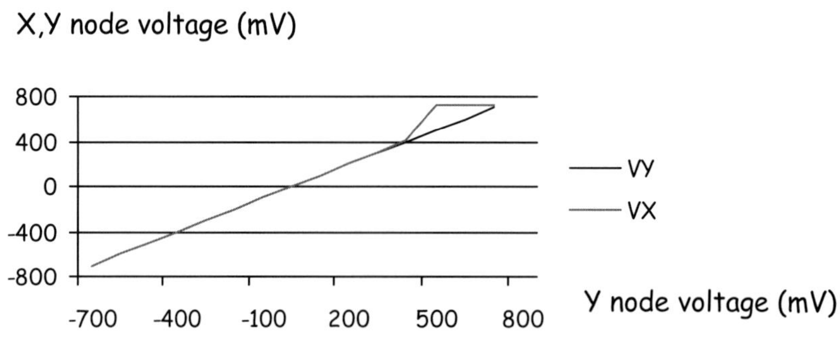

Figure B.4 – Dynamic range of figure B.1 topology

Then, the current characteristic has been measured, so verifying that the current flowing from X node is equal to that flowing from Z node. Two 10 kΩ resistances have been connected to X and Z terminals and the current has been measured for different values of the input voltage V_Y. Experimental results are shown in figure B.5.

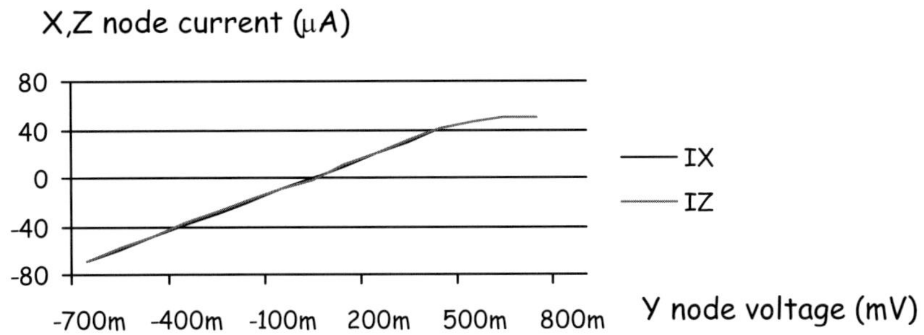

Figure B.5 – Current characteristic of figure B.1 topology

Simulation and measurement results on this CCII are summarised in table B.1.

Current Conveyor Characteristics		
Parameter	**Simulated Value**	**Measured Value**
Voltage Gain (α)	1.0001	1.00
Current Gain (β)	1.00	1.00 (Rx=Rz=10KΩ)
3dB Bandwidth	37 MHz	3.2 MHz
Dynamic Range	–600mV, +500mV	–750mV, +400mV
Node X Parasitic Resistance	7 Ω	23 Ω
Node Z Parasitic Impedance	190 kΩ	~10 MΩ

Table B.1 – CCII characteristics for the circuit shown in figure B.1

The behaviour of the fabricated current conveyor is quite good, in particular the dynamic range does not seem to be smaller even with low resistive loads.
In order to confirm the measure results, the circuit performance have been evaluated through a digital oscilloscope applying to the Y node a sinusoidal signal having a frequency of 10 kHz and a variable amplitude. This amplitude has been incremented up to the maximum value to which corresponds, at X node, an undistorted signal. In figure B.6 the upper trace is the applied Y node voltage, while the lower one is the corresponding voltage at X node.
It is possible to note that the X node voltage shows an undistorted peak-to-peak amplitude of about 725 mV, thus confirming the graph shown in figure B.4.

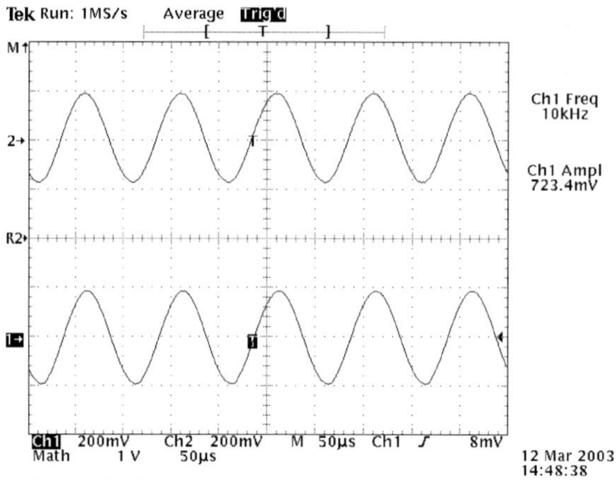

Figure B.6 – Transient response of figure B.1 topology
(upper trace: input voltage – lower trace: output voltage)

Once the maximum amplitude have been reached, a second measure has been performed applying to Y node a signal having an amplitude of 750 mV (that is 1.5 V peak-to-peak), whose value is equal to the voltage supply. The input signal and the corresponding X node signal are presented in figure B.7.
Obviously, the output signal is extremely distorted due to output saturation, even if its swing is particularly high.

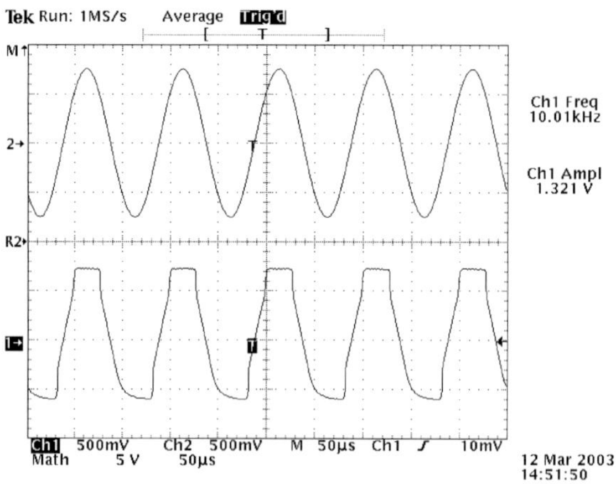

Figure B.7 – Transient response of figure B.1 topology with maximum dynamic input signal
(upper trace: input voltage – lower trace: output voltage)

As stated in chapter 2, the current conveyor presented in figure B.1 has a serious drawback represented by its output stage, whose biasing current cannot be controlled easily since its value depends dramatically on process variations and supply voltage. A step towards the solution of such a problem is represented by the circuit shown in figure B.8, where the biasing current in transistors M5 and M6 is controlled by means of I_{BIAS2} current and R resistance.

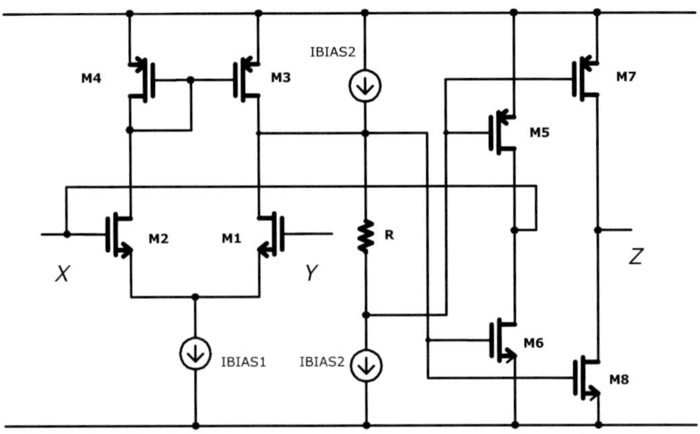

Figure B.8 – Second CCII topology fabricated

In figure B.9, a microphotograph of this CCII is presented.

Figure B.9 – Microphotograph of figure B.8 topology

As seen for the first topology, the frequency response at X node is evaluated. Also in this case, experimental results show a peak in the frequency response and the bandwidth seems to be a little bit smaller with respect to the one of the first CCII presented in this appendix. The data are summarised in the plot shown in figure B.10.

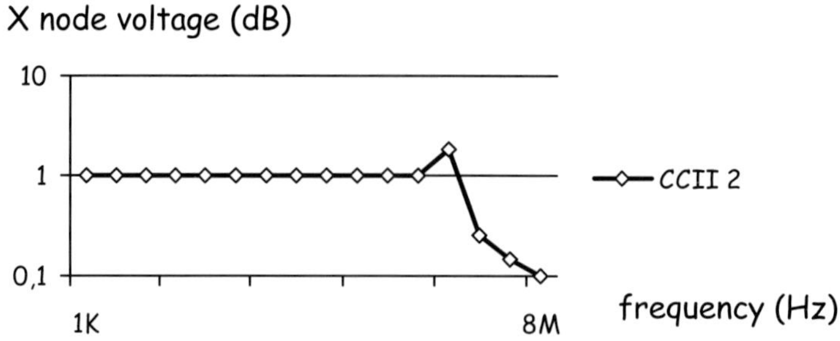

Figure B.10 – Measured frequency response of figure B.8 topology

Then, the dynamic range is investigated, whose experimental results are presented in figure B.11. Analysing the topology implemented (see figure B.8) the behaviour for negative input voltages is very close to the expected one, while, for positive input voltages, the corresponding voltage at X node saturates to the positive supply. This means that during chip fabrication some MOS parameters have varied so affecting some CCII performance.

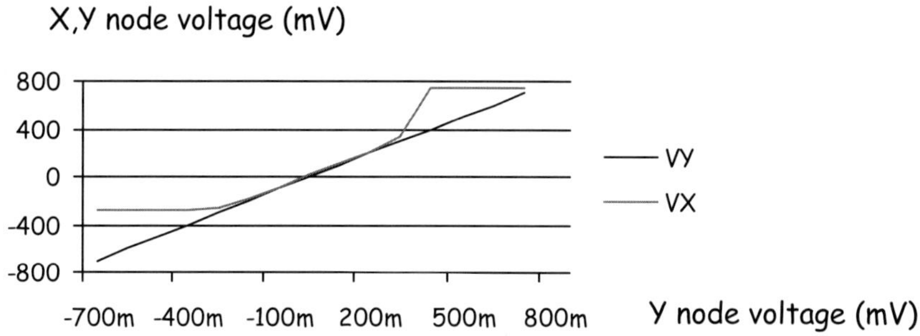

Figure B.11 – Dynamic range of figure B.8 topology

Connecting two load resistances at X and Y nodes, the current characteristic has been evaluated, whose results (I_X and I_Z vs. V_Y) are summarised in figure B.12 graph.

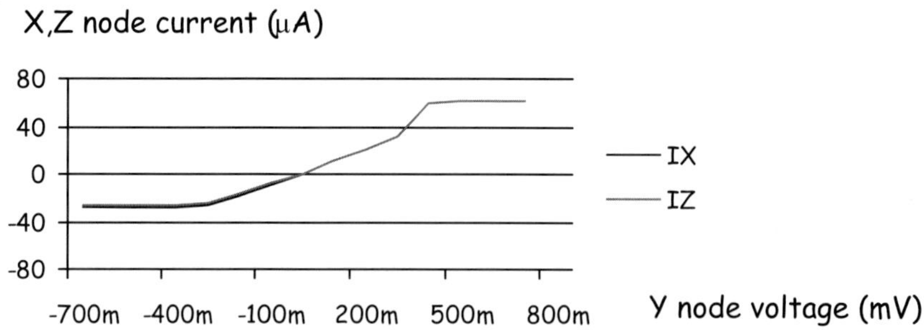

Figure B.12 – Current characteristic of figure B.8 topology

The dynamic range seems to be improved, but this is only due to the particular value of load connected. The complete simulation and measurement results are reported in table B.2.

Current Conveyor Characteristics		
Parameter	**Simulated Value**	**Measured Value**
Voltage Gain (α)	1.0006	1.00
Current Gain (β)	1.00	1.00 (Rx=Rz=10KΩ)
3dB Bandwidth	15.3 MHz	1.45 MHz
Dynamic Range	–400mV, +450mV	–250mV, +250mV
Node X Parasitic Resistance	26 Ω	92 Ω
Node Z Parasitic Impedance	295 kΩ	~11 MΩ

Table B.2 – CCII characteristics for the circuit shown in figure B.8

Measurements with the digital oscilloscope confirm that the dynamic range is limited, for 10 KHz frequency, to signals having amplitude of about 200 mV (or 400 mV peak-to-peak), as shown in figure B.13.

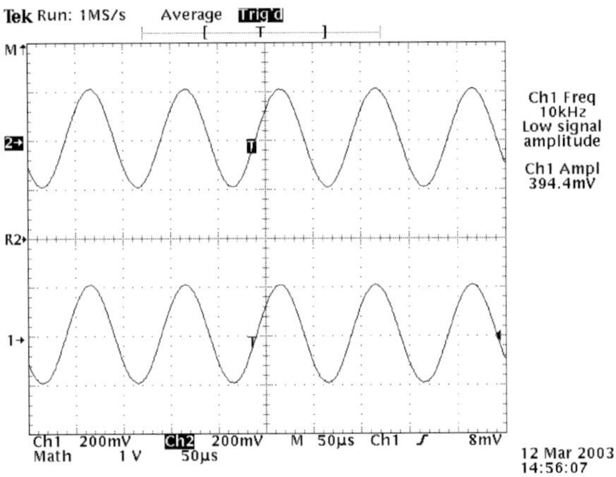

Figure B.13– Transient response of figure B.8 topology (upper trace: input voltage – lower trace: output voltage)

Imposing an input signal of 1.5 V peak-to-peak, the corresponding output signal at X node is reported in figure B.14. The output stage clearly saturates and the signal has a swing of about 1 V, thus confirming the graph pictured in figure B.11.

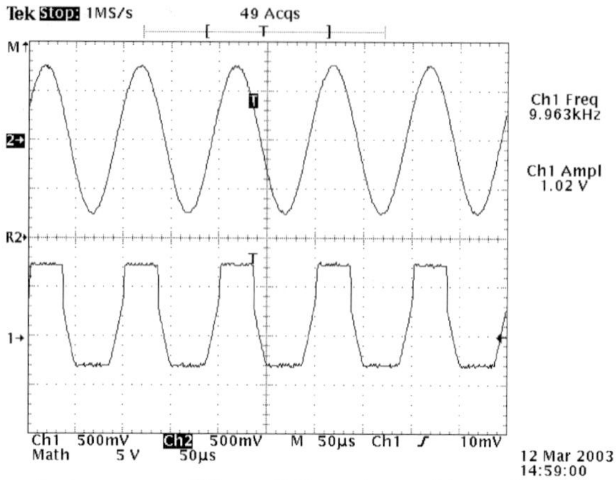

Figure B.14 – Transient response of figure B.8 topology with maximum dynamic input signal (upper trace: input voltage – lower trace: output voltage)

The complementary topology, using a p-type differential pair, has been also fabricated and is shown in figure B.15, while its microphotograph is depicted in figure B.16.

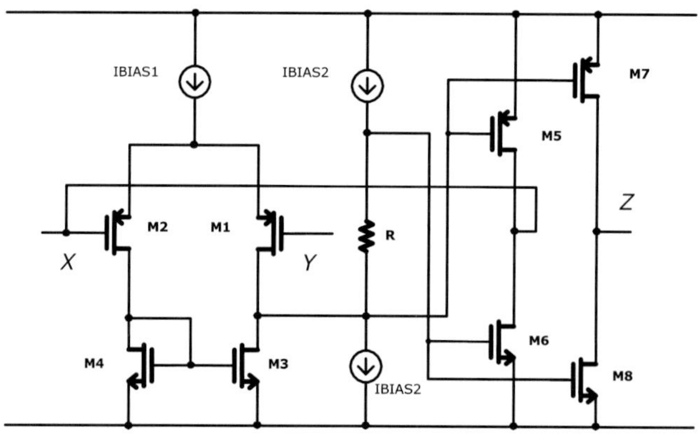

Figure B.15 – Third CCII topology fabricated

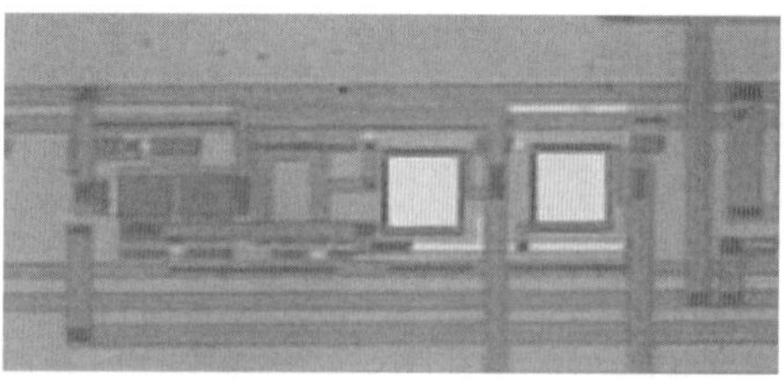

Figure B.16 – Microphotograph of figure B.15 topology

The frequency response of the third CCII fabricated is reported in figure B.17. Comparing to the other two topologies presented before, the bandwidth is reduced, even if it seems better compensated.

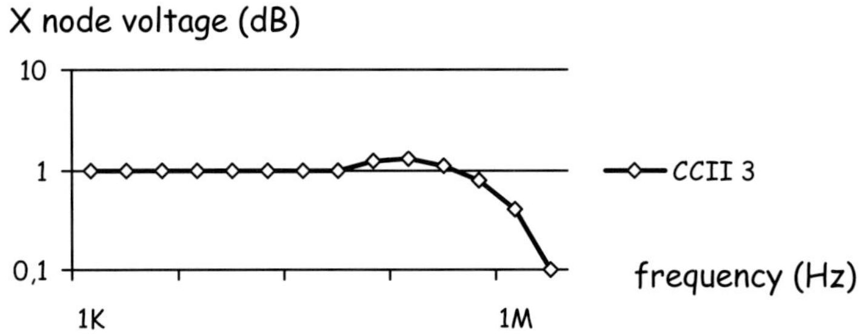

Figure B.17 – Measured frequency response of figure B.15 topology

The measured dynamic range, presented in figure B.18, is in an excellent agreement with the theory and the simulations.

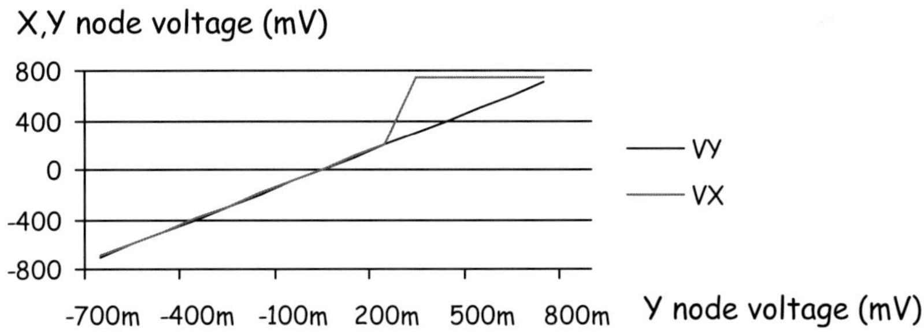

Figure B.18 – Dynamic range of figure B.15 topology

The measure of the current characteristic (I_X and I_Z vs. V_Y) shows a little reduction in the operating range for what concerns the positive input voltages, while for negative inputs the CCII operates correctly.

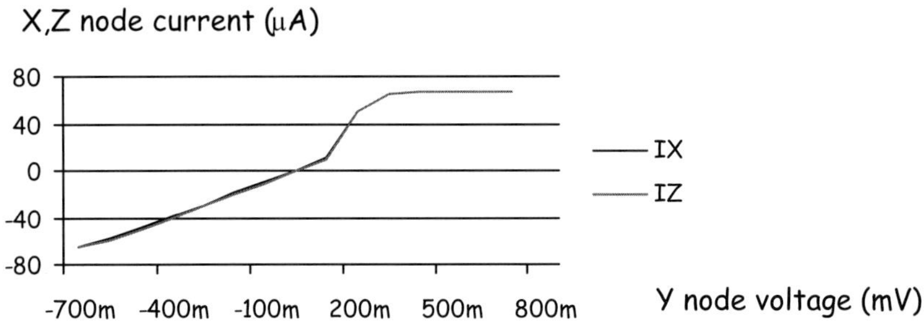

Figure B.19 – Current characteristic for CCII 3

Also for this topology the simulated and measured CCII characteristics have been summarised in a table (table B.3).

Current Conveyor Characteristics		
Parameter	**Simulated Value**	**Measured Value**
Voltage Gain (α)	1.0003	1.00
Current Gain (β)	1.00	1.00 (Rx=Rz=10KΩ)
3dB Bandwidth	18 MHz	650 kHz
Dynamic Range	–750mV, +350mV	–750mV, +250mV
Node X Parasitic Resistance	34 Ω	87 Ω
Node Z Parasitic Impedance	323 kΩ	~5.3 MΩ

Table B.3 – CCII characteristics for the circuit shown in figure B.15

In agreement with the dynamic response shown in figure B.18, the maximum input signal, at 10 KHz frequency, that gives an undistorted output at X node, has an amplitude of about 100 mV (200 mV peak-to-peak), as shown in figure B.20.

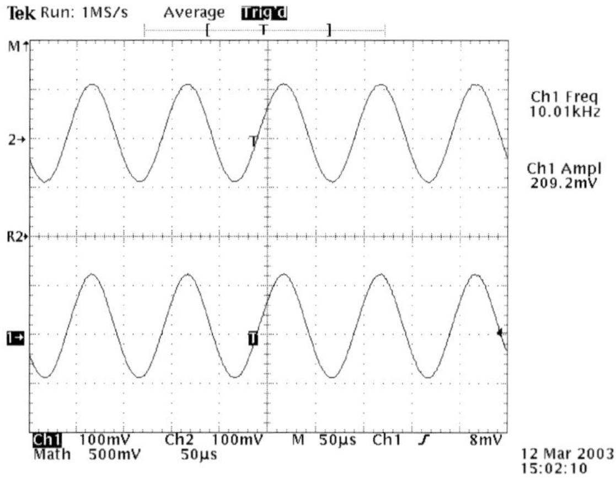

Figure B.20– Transient response of figure B.15 topology (upper trace: input voltage – lower trace: output voltage)

Imposing a 1.5 V peak-to-peak input signal, we obtain the output voltage at X node whose shape is pictured in figure B.21 (lower trace). It is interesting to note that output stage saturation occurs only for the positive half wave, while the negative one follows the ideal behaviour very well. This is in an excellent agreement with the characteristic shown in figure B.18 and it is due to the fact that for negative input voltages the transistors of the differential pair are ON, ensuring a correct operating mode. When the input voltage increases, the CCII stops to operate correctly.

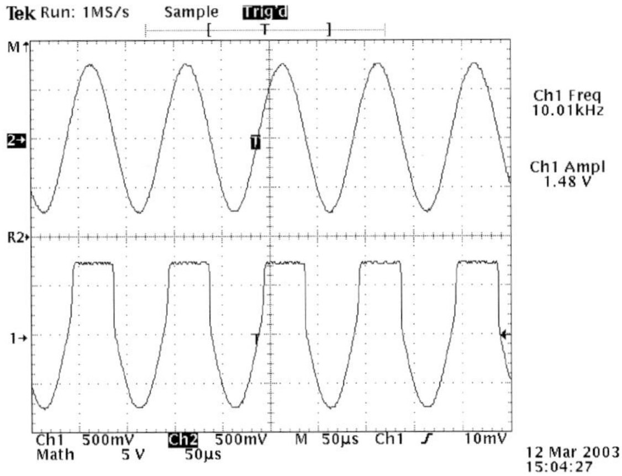

Figure B.21 – Transient response of figure B.15 topology with maximum dynamic input signal (upper trace: input voltage – lower trace: output voltage)

The rail-to-rail version of the CCIIs shown in figures B.8 and B.15 implements both n-type and p-type input differential pairs, as in figure B.22. The chip microphotograph is shown in figure B.23.

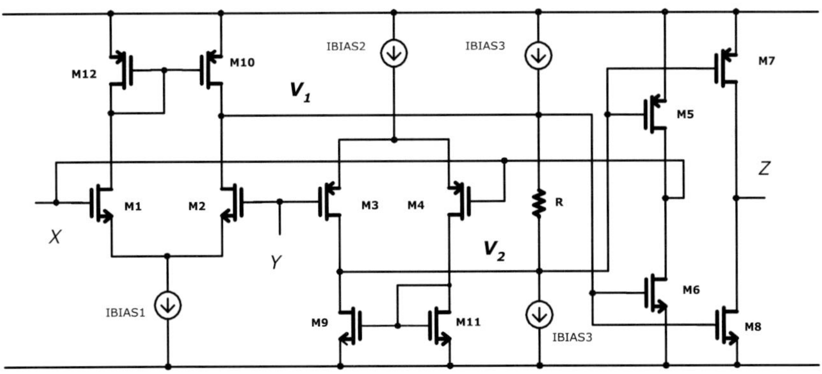

Figure B.22 – Fourth CCII topology fabricated

Figure B.23 – Microphotograph of figure B.22 topology

The frequency response, presented in figure B.24, shows once more the peak at MHz frequencies, even if the cut-off frequency is now higher with respect to that measured for the topologies previously described.

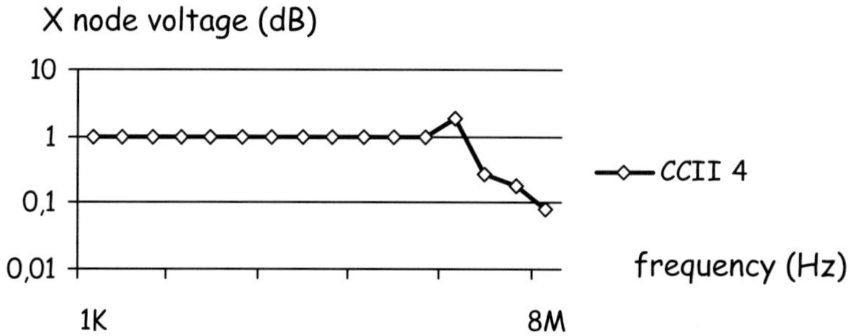

Figure B.24 – Measured frequency response of figure B.22 topology

The circuit shows a full dynamic range, as shown in figure B.25.

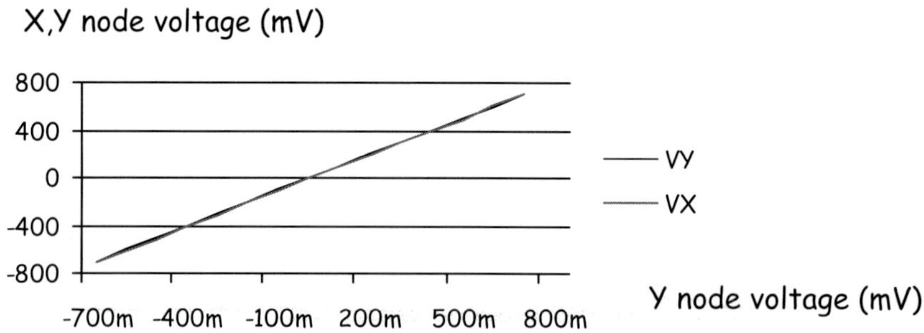

Figure B.25 – Dynamic range of figure B.22 topology

Figure B.26 shows the DC current characteristics (I_X and I_Z vs.V_Y) for this CCII, showing good performance.

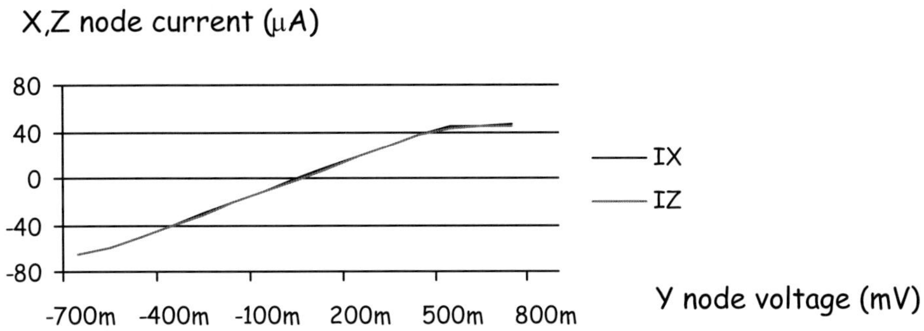

Figure B.26 – Current characteristic of figure B.22 topology

Table B.4 summarises simulation results and experimental measures of the CCII characteristics.

Current Conveyor Characteristics		
Parameter	**Simulated Value**	**Measured Value**
Voltage Gain (α)	1.00006	1.00
Current Gain (β)	1.00	1.00 (Rx=Rz=10KΩ)
3dB Bandwidth	40 MHz	1.85 MHz
Dynamic Range	–750mV, +650mV	–750mV, +500mV
Node X Parasitic Resistance	24 Ω	50 Ω
Node Z Parasitic Impedance	324 kΩ	~3.8 MΩ

Table B.4 – CCII characteristics for the circuit shown in figure B.22

For this circuit we have applied an input voltage of 1 V amplitude and 10 KHz frequency (see figure B.27), obtaining an undistorted output signal. This confirms the large dynamic range of this configuration.
Figure B.28 shows the output response to a 1.5 V peak-to-peak input signal. It is interesting to note that a rail-to-rail operating range is not possible, due to the distortion of the positive half-wave of the input signal. This situation confirms theory and simulations.

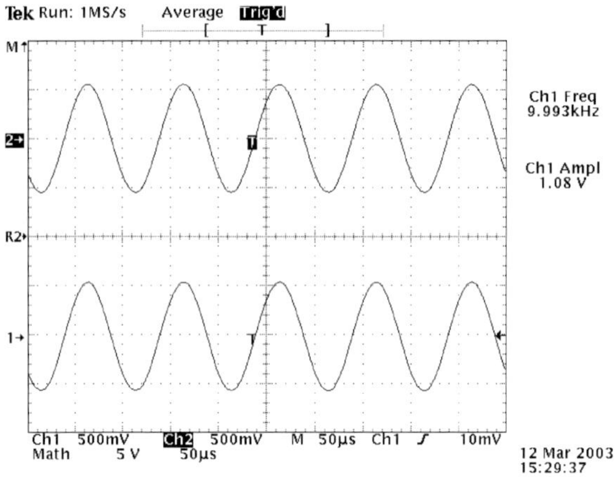

Figure B.27– Transient response of figure B.22 topology
(upper trace: input voltage – lower trace: output voltage)

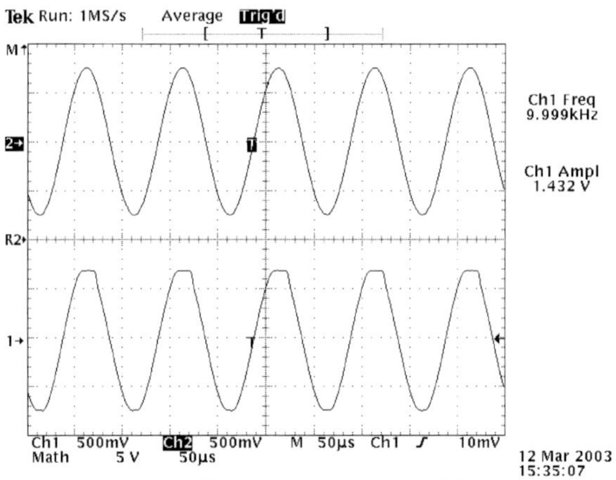

Figure B.28 – Transient response of figure B.22 topology with maximum dynamic input signal
(upper trace: input voltage – lower trace: output voltage)

The last fabricated topology of CCII is reported in figure B.29. Figure B.30 shows its microphotograph.

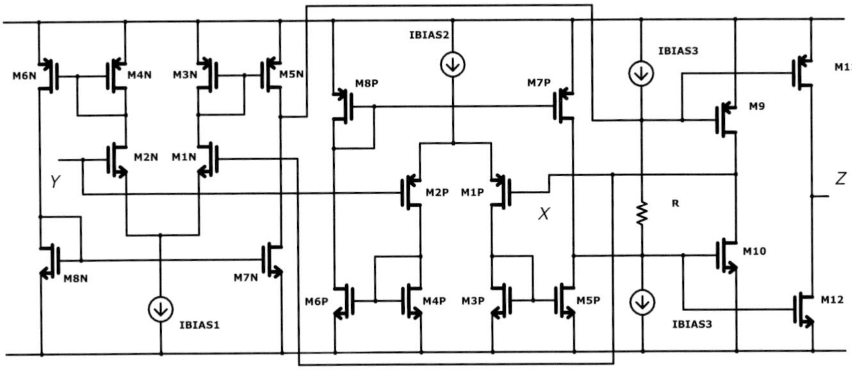

Figure B.29 – Fifth CCII topology fabricated

The CCII shown in figure B.29 is based on two symmetrical OTAs and, as reported in the previous chapters, ensures a true rail-to-rail behaviour.

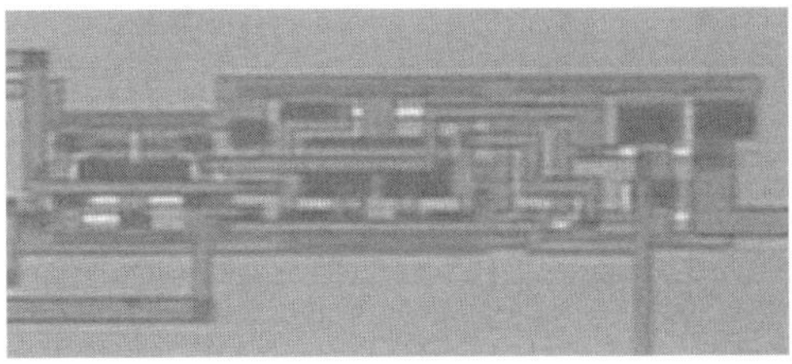

Figure B.30 – Microphotograph of figure B.29 topology

In this case the output peak in the frequency response is present at very high frequencies (see figure B.31).

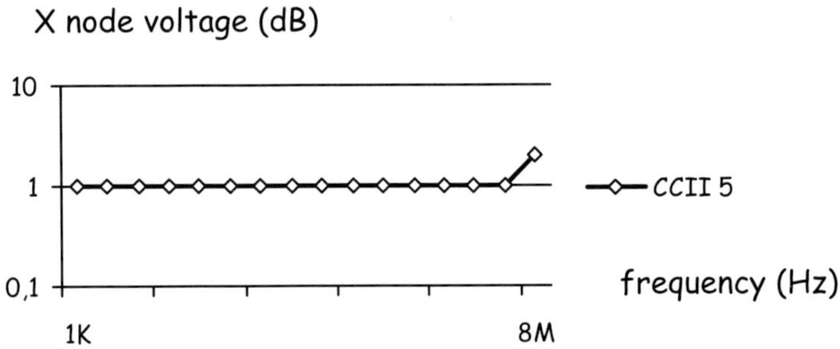

Figure B.31– Measured frequency response of figure B.29 topology

The dynamic range is, as expected, very large (see figure B.32), even if it is not completely rail-to-rail.

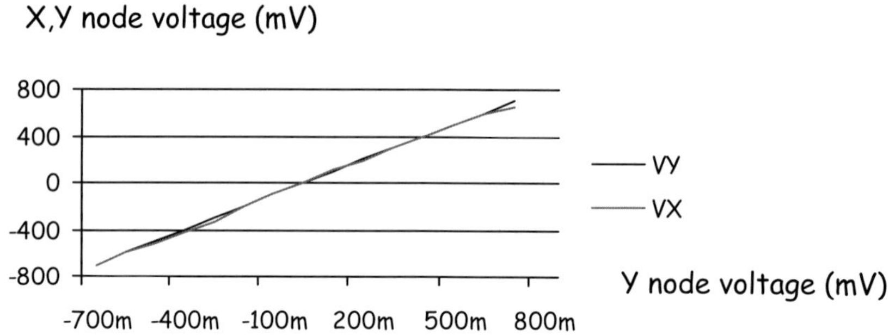

Figure B.32 – Dynamic range of figure B.29 topology

Figure B.33 shows the current characteristic, which is almost rail-to-rail.

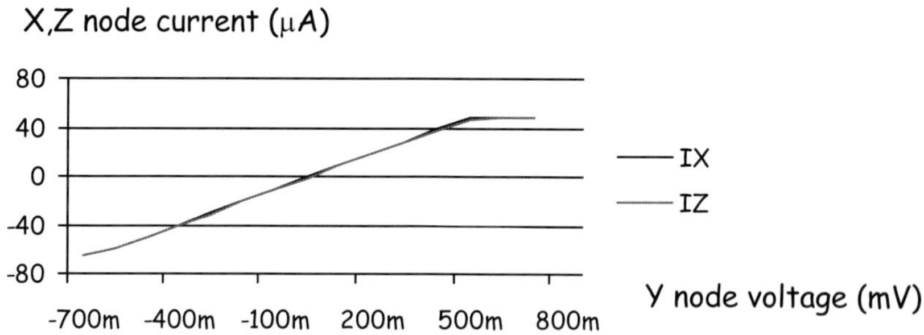

Figure B.33 – Current characteristic of figure B.29 topology

Simulation and measurement results are summarised in table B.5.

Current Conveyor Characteristics		
Parameter	**Simulated Value**	**Measured Value**
Voltage Gain (α)	1.00018	1.00
Current Gain (β)	1.00	1.00 (Rx=Rz=10KΩ)
3dB Bandwidth	55 MHz	15 MHz
Dynamic Range	–750mV, +750mV	–750mV, +600mV
Node X Parasitic Resistance	3.5 Ω	147 Ω
Node Z Parasitic Impedance	79 kΩ	~4.8 MΩ

Table B.5 – CCII characteristics for the circuit shown in figure B.29

Measurements with digital oscilloscope confirm (see figure B.34) that the maximum input voltage amplitude is about 600 mV (or 1.2 V peak-to-peak). In fact, a little distortion has been measured in the output signal when the input one has a 1.5 V peak-to-peak amplitude, as shown in figure B.35.

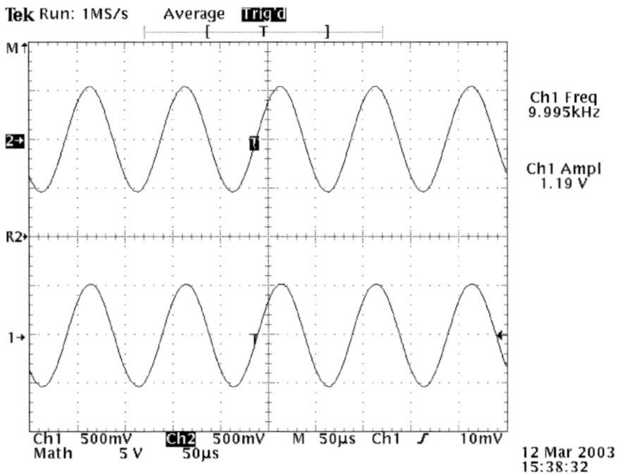

Figure B.34– Transient response of figure B.29 topology (upper trace: input voltage – lower trace: output voltage)

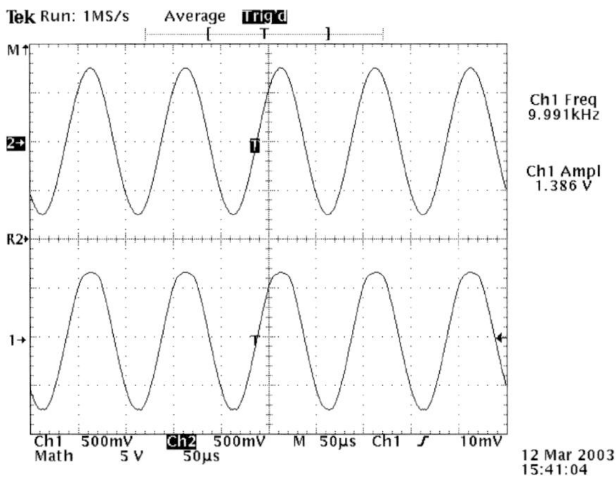

Figure B.35 – Transient response of figure B.29 topology with maximum dynamic input signal (upper trace: input voltage – lower trace: output voltage)

Appendix C

In chapters 2 and 3 several possible implementations for integrated current conveyors have been presented. In particular, we have focused our attention on the LV LP solutions.

If there is not a particular need of LV LP topologies (or it is not possible to design and fabricate a chip dedicated to a CCII-based application), a component commercially available can be the best solution. Such a component is the Analog Device AD844, whose simplified block scheme is shown in figure C.1.

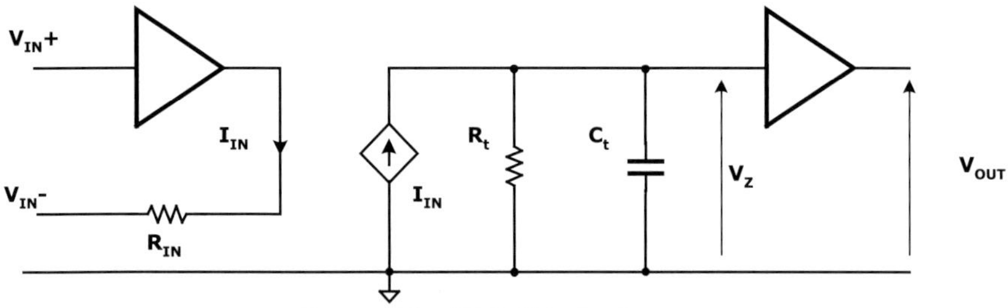

Figure C.1 – AD844 block scheme

The input stage is represented by a voltage buffer and a R_{IN} resistance. The node V_{IN+} shows a high impedance, while V_{IN-} shows a low one, ideally zero. When a voltage is applied to V_{IN+} and a load is connected to V_{IN-} , a current named I_{IN} flows in this load. This current can be found also at the output stage, represented as the current source I_{IN}. The load connected to this current source represents the transresistance of the AD844 amplifier. The ratio between R_t and R_{IN} is the voltage gain, having a typical value of 60000, while current gain is about 40000. At the output we have two nodes: a high impedance node (Z) and a low impedance one (OUT). According to the theory presented in the first chapter, this device can be considered as a Current Feedback Operational Amplifier (CFOA). It is interesting to remind that a CFOA can be viewed as a current conveyor followed by a voltage buffer or as two current conveyors. Moreover, considering, in figure C.1, only the terminals V_{IN+}, V_{IN-} and Z, the AD844 operates as a current conveyor.

Minimum, typical and maximum values of the main characteristics of the AD844 have been reported in table C.1. More detailed information can be found in the datasheet of the component, available on the web-site of the Analog Devices.

Parameter	Value			Unit
	MIN	TYP	MAX	
Input offset voltage		50	300	μV
Input bias current+		200	450	nA
Input bias current–		150	400	nA
Input resistance+	7	10		MΩ
Input resistance–		50	65	Ω
Open loop transresistance	2.2	3.0		MΩ
Output resistance		15		Ω
Supply voltage	4.5		18	V

Table C.1 – AD844 main characteristics (from the datasheet)

In a wide number of applications the AD844 allows a quick implementation of different CCII-based solutions. As an example, we have considered the floating inductance simulator, presented in chapter 5 (see figure 5.27) and here pictured again in figure C.2.

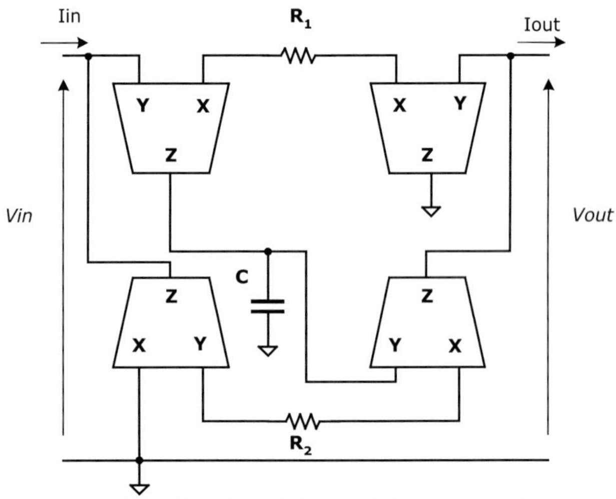

Figure C.2 – CCII-based floating inductance simulator

To evaluate this solution, which allows to obtain an equivalent inductance valued R_1R_2C, a board, shown in figure C.3, has been fabricated.

Figure C.3– Photograph of the CCII-based floating inductance simulator using AD844

The circuit has been tested to obtain very large inductance values without compromising the dynamic range. In figure C.4 the different inductance values obtained with different capacitances have been reported. The experimental results are very satisfactory.

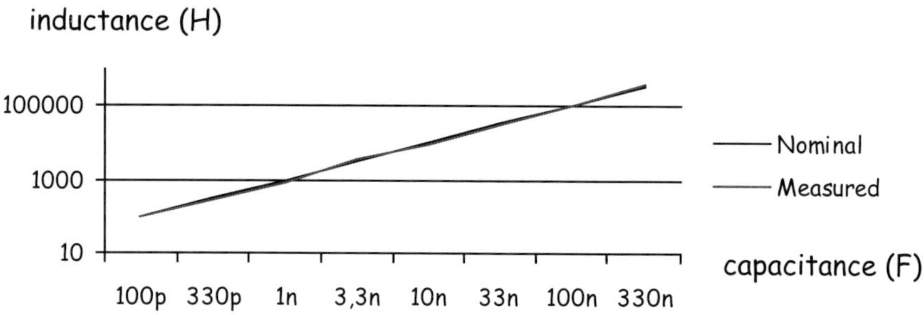

Figure C.4 – Inductance vs. capacitance C value

A variation of the capacitance value implies a corresponding variation of the operating frequency bandwidth. In fact, lowering the capacitance value its impedance (at a fixed frequency) increases. This means that it can be a problem to drive small capacitances, especially at low frequencies. That's why if the simulated inductance is needed for very low frequency applications, the capacitance value can not be chosen too small.

The simulated inductance is of a floating type, so it is possible to use it as a series element in electrical circuits. A very simple example is represented by a low pass LR filter. In figure C.5, input and output signals, at the cut-off frequency, are reported. Resistances R_1 and R_2 and capacitance C have nominal values of 10 KΩ, 100 KΩ and 330 pF, respectively. The expected inductance value (307 mH) has been evaluated considering the measured component values, that are, for R_1 and R_2 and C, 9.89 KΩ, 99.7 KΩ and 312 pF, respectively. The measured value of 303 mH confirms the validity of the presented solution and the accuracy of the results.

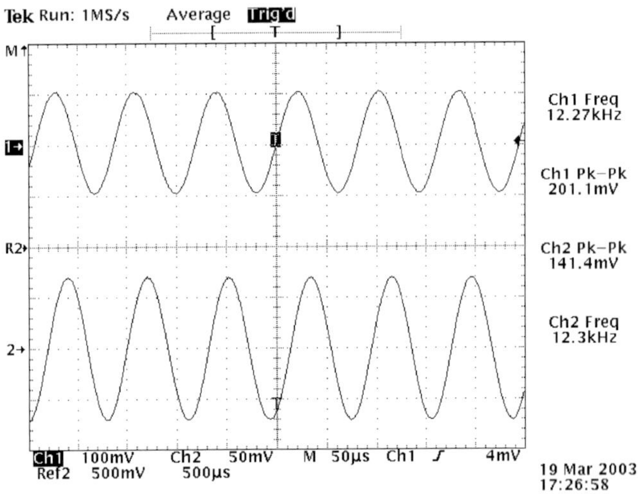

Figure C.5 – Time response of the L-R filter implemented
(upper trace: input voltage – lower trace: output voltage)